Analog Circuits and Signal Processing

Series Editors

Mohammed Ismail, Dublin, USA
Mohamad Sawan, Montreal, Canada

More information about this series at http://www.springer.com/series/7381

Marco Vigilante · Patrick Reynaert

5G and E-Band Communication Circuits in Deep-Scaled CMOS

Springer

Marco Vigilante
ESAT-MICAS
KU Leuven
Leuven
Belgium

Patrick Reynaert
ESAT-MICAS
KU Leuven
Leuven
Belgium

ISSN 1872-082X ISSN 2197-1854 (electronic)
Analog Circuits and Signal Processing
ISBN 978-3-319-72645-8 ISBN 978-3-319-72646-5 (eBook)
https://doi.org/10.1007/978-3-319-72646-5

Library of Congress Control Number: 2017964595

Printed on acid-free paper

This Springer imprint is published by the registered company Springer International Publishing AG part of Springer Nature
The registered company address is: Gewerbestrasse 11, 6330 Cham, Switzerland

Preface

We are at the dawn of a new era. New emerging applications will revolutionize the way we communicate, share ideas, work, travel, play, watch sports, and enjoy movies; in a single word, the way we live. For Internet of Things (IoT) applications, it is estimated that up to hundred devices will be connected and share information for each person, from wearable devices -(such as smartwatches) to disposable lab-on-a-chip (for smart health care). Those devices will generate an enormous amount of data, posing unprecedented challenges on each element of the network. For automotive applications, advanced driver assistance systems (ADAS) are expected to evolve in self-driving cars with automatic parking and predictive-collision-avoidance features. For mobile applications, virtual reality (VR) games and videos are expected in the near future.

Fifth generation mobile networks (5G) is the wireless standard that will address these challenges. 100× higher data rate is needed at 100× higher network efficiency. For the network to provide high-quality services such as 3D 360° video and 360° surround solutions to enable virtual reality (VR) while being transparent to the user, better than 1 ms latency is needed. This is the first time that a wireless standard puts such stringent specifications to improve the user experience. To send so much data in such a limited time, an enormous amount of bandwidth is required. The spectrum in the low GHz range is already overcrowded; therefore, mm-Wave wireless communication is going to happen in the near future.

CMOS is the technology of choice for mass production digital circuits. It guarantees high yield and low costs, while the aggressive scaling of the minimum feature size allows to integrate low power mm-Wave analog building blocks together with the baseband digital signal processing. CMOS is therefore a key technology for the success of 5G mm-Wave front-ends and has attracted a growing attention in the last decade from both industries and research institutes.

However, aggressive technology scaling does not provide only benefits. The low-level metal interconnections get thinner and closer to the substrate, seriously limiting the achievable f_{MAX} of active devices and the maximum quality factor of on-chip passive devices. The supply voltage scales as well, making the classical analog design trade-offs tighter. Moreover, the requirements on large bandwidth of operation should be met under process, voltage, and temperature (PVT) variations, and extra margin should be taken to allow substantial model inaccuracy due to the high frequency of operation.

This work focuses on these challenges and proposes design techniques for several building blocks that currently limit the performance of mm-Wave transceivers. The distinctive features of high-speed analog design in deep-scaled CMOS will be addressed, and a comparison with older technology node will be provided. Transformer-based low loss broadband filters that realize interstage matching, power division/combining, and impedance transformation will be discussed in great detail. Simple design equations that shed new insights on these pervasive kinds of filters will be provided. Second-order effects due to physical layout implementation will be addressed and simple solutions will be proposed. Tuning extension techniques for integrated mm-Wave oscillators will be discussed. The design, layout, and measurements details of five state-of-the-art building blocks that leverage the proposed design techniques will be presented. (1) An E-Band quadrature voltage controlled oscillator tunable over two bands of almost 5 GHz each separated in frequency, while achieving state-of-the-art phase noise and power consumption is demonstrated. The integrated prototype realizes accurate quadrature phases and occupies only 0.031 mm^2. (2) A wideband inductor-less frequency divide-by-4 that allows low power operation with wide margin over the whole E-Band (60−90 GHz) and beyond is reported for the first time. (3) A broadband low-noise amplifier for E-Band point-to-point communication links that achieves a figure of merit $\approx$10.5 dB better than the state-of-the-art designs in the same band is shown. (4) The LNA is further integrated into a broadband sliding-IF receiver that demonstrates 30.8 dB conversion gain with ¡1 dB in-band ripple over a 27.5 GHz BW_{-3dB} while achieving a 7.3 dB minimum NF with less than 2 dB variation from 61.4 to 88.9 GHz. This wideband state-of-the-art performance enables robust and low power multi-Gb/s wireless communication over short to medium distance over the complete E-Band with wide margin. (5) A 29–57 GHz (65% *BW*) AM-PM compensated class-AB power amplifier tailored for 5G phased arrays is demonstrated. This integrated prototype shows outstanding AM-PM linearity allowing excellent EVM and ACPR while amplifying wideband modulated signals with high PAPR. All designs were implemented in a 28-nm CMOS technology without RF ultra-thick top metal option.

Leuven, Belgium
October 2017

Marco Vigilante
Patrick Reynaert

Contents

Chapter 1
Introduction

1.1 Towards 5G and IoT

The evolution of mobile communication has a deep impact on the daily life of millions of people all over the world. In just a few decades, we have witnessed a revolution in the way people communicate, share ideas and live. This is still happening and will continue in the future. The 1G analog cellular system was introduced in the '80s. But it is only with the 2nd generation 2G and the switch to digital cellular system that in the '90s the mobile communication reached the mass level production, connecting people all over the world. Today, thanks to 3G ('00s) and 4G ('10s) people are able to use mobile devices to connect to the internet. This phenomenon is referred as people-to-thing communication.

Internet of Things (IoT) is happening next, aiming to connect people and objects everywhere at anytime. 5G will be the key enabler of the IoT and, following the trend of the previous generations, its full deployment is expected in 2020 [1]. Figure 1.1 shows the requirements for such technology and compares them to 4G. Together with the classical requirements of higher data rate and spectrum efficiency, connection density, area traffic capacity and latency are becoming key features. These requirements are fundamental to improve the user experience, core added value of the IoT. Moreover, these specifications should be met while achieving 100× better network efficiency [2]. 5G will enable safer transportation, better healthcare and smart objects, improving further our quality of life.

IoT therefore needs a low cost and low power technology, so that every object around us can become smart while requiring a small battery or no battery at all. CMOS is the technology of choice for mass production digital circuits for these same reasons. Since more than 50 years CMOS technology scaling has followed the so called Moores law. Every new generation allows to integrate more transistors (therefore more functions) in the same area (hence at the same cost), with a reduced power consumption. CMOS therefore is playing a key role in the IoT [4].

M. Vigilante and P. Reynaert, *5G and E-Band Communication Circuits in Deep-Scaled CMOS*, Analog Circuits and Signal Processing,
https://doi.org/10.1007/978-3-319-72646-5_1

Together with this win-win relationship between lower cost and lower power, at each technology node the MOS transistors get faster. Digital processors therefore enjoy the full benefit of technology scaling, but what about analog design in advanced CMOS?

1.2 mm-Wave Spectrum, Challenges and Opportunities

To achieve the performance summarized in Fig. 1.1, 5G needs to be a leap forward from 4G. This can not be possibly accomplished by a simple incremental advance on previous technologies. Shannon in '48 [5] demonstrated that the fundamental limit of the channel capacity (C) is proportional to the channel bandwidth (BW):

$$C = BW\, log_2(1 + SNR), \tag{1.1}$$

where SNR is the signal-to-noise ratio. This is one of the fundamental reasons why industries and research institutes are pushing towards solutions at higher frequencies, where more bandwidth is available. However, the need for higher frequency has to face unprecedented challenges. The attenuation that a transmitted signal undergoes in free space (known as free space path loss, FSPL) is expressed as

$$FSPL = \left(\frac{4\pi\, df}{c}\right)^2, \tag{1.2}$$

where d is the distance, f the frequency, and c the speed of light. The higher the frequency the higher the loss. Moreover, the signal will propagate through air (and not in free space). The resulting attenuation at sea level is shown in Fig. 1.2. The oxygen O_2 present in the atmosphere causes a clear peak at 60 GHz, followed by an

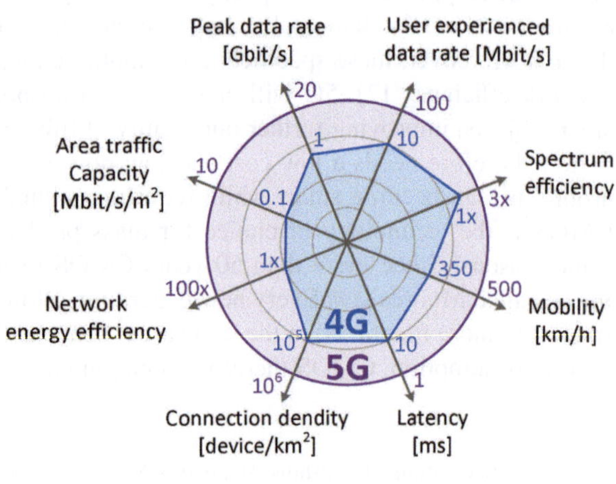

Fig. 1.1 5G requirements and comparison against 4G [2]. ©2017 John Wiley and Sons. Reprinted, with permission, from [3]

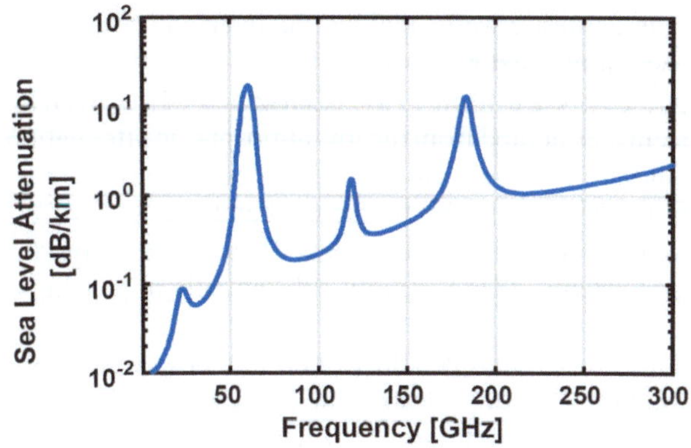

Fig. 1.2 Sea level attenuation against frequency. ©2017 John Wiley and Sons. Reprinted, with permission, from [3]

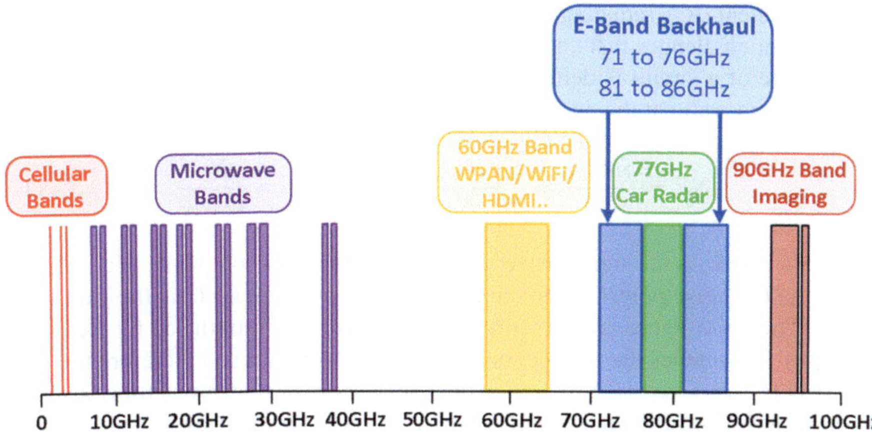

Fig. 1.3 Major spectrum allocation in the United States. ©2017 John Wiley and Sons. Reprinted, with permission, from [3]

atmospheric window between 70 and 90 GHz. The different propagation characteristics of the medium are the reason for the allocation of specific services in particular parts of the spectrum.

If on the one hand 5G will be back compatible to 3G and 4G, and will benefit from similar technology and solutions. On the other, to push further the data rate, mm-Wave frequencies will be a key evolution with respect to the previous technologies [1]. Figure 1.3 shows that a channel from 57 to 66 GHz is reserved to high speed short range wireless communications. The high atmospheric absorption of ≈12 dB/km permits the coexistence of several different WPAN, WiFi and HDMI services in a small area. Such wireless personal area networks would not be able to extend their signals through the domestic walls, making the interferer to a network operating at the same frequency in the next room negligible.

Two bands of 5 GHz each from 71 to 76 GHz and 81 to 86 GHz are reserved to backhauling systems. Benefited from the low atmospheric attenuation (<0.5 dB/km), such systems could provide multi Gb/s links for fiber extension or replacement over short to medium distances [6]. The frequency band that spans from 77 to 81 GHz has been allocated for car radar applications. These radars would make Advanced Driver Assistance Systems (ADAS) reality, substantially improving the safety on our roads [7].

All these applications would benefit from a low power and low cost fully integrated CMOS solution. However, despite the aggressive technology scaling, severe challenges are posed on the high frequency analog front-end. Figure 1.4 shows the cut-off frequency against minimum channel length [8]. Indeed, every technology node shows a clear advantage in speed. Even if in deep sub-micron technology the effect of velocity saturation becomes dominant also for moderate values of the Inversion Coefficient (IC), reducing the slope from 20 to 10 dB/dec. Figure 1.5 shows two of the major challenges that an analog designer faces at mm-Wave frequencies. The noise figure of a circuit is defined as the signal-to-noise ratio at the input over the signal-to-noise ratio at the output [9]

$$NF = \frac{SNR_{IN}}{SNR_O} = \frac{S_{IN}/N_{IN}}{S_O/N_O} = \frac{S_{IN}\,N_O}{G\,S_{IN}\,N_{IN}} = \frac{N_O}{G\,N_{IN}}, \tag{1.3}$$

where S_{IN} and S_O is the signal power at the circuit input and output respectively, N_{IN} and N_O is the noise power at the circuit input and output, and G is the gain of the circuit. The noise figure is a measure of the excess noise introduced by the circuit. At higher frequencies the transconductor shows lower gain and therefore the noise noise figure rises (see Fig. 1.5). So, on the one hand transistors are getting faster, but on the other the performance degradation at higher frequencies will have a serious impact on circuit design. Moreover, to ensure reliability while the minimum channel length aggressively scales, the supply voltage needs to follow. This trend is clearly visible in Fig. 1.6. The implications of this phenomenon will be deeply discussed in the following chapters. In this section we limit our discussion to the following. (1) The phase noise in a VCO is relative to the carrier power, which in turns is proportional to the supply voltage (for any oscillator topology). (2) The maximum output power that a power amplifier is able to deliver, is also proportional to V_{DD} (for any PA topology). (3) The number of devices that can be stacked to realize a cascode amplifier and/or a current source is limited by V_{DD} and V_t, and the latter can not scale as much (see Fig. 1.6).

Finally, the higher the frequency the smaller the feature size of the antenna. The mm-Wave spectrum allows therefore not only the use of on-chip antenna, but also antenna arrays with a large number of elements, making massive MIMO and beamforming key technologies for 5G [1].

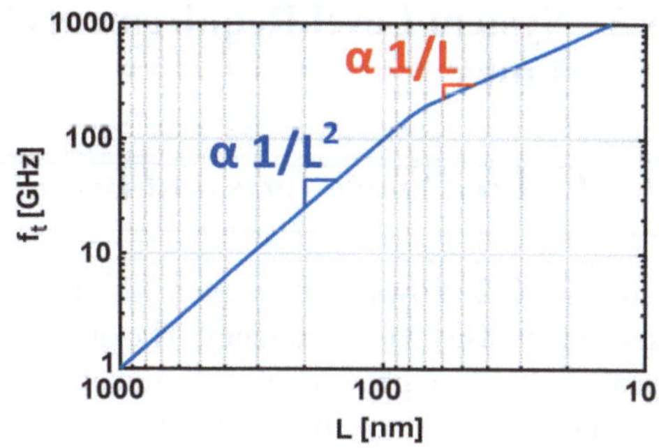

Fig. 1.4 Cut-off frequency against channel length [8]. ©2017 John Wiley and Sons. Reprinted, with permission, from [3]

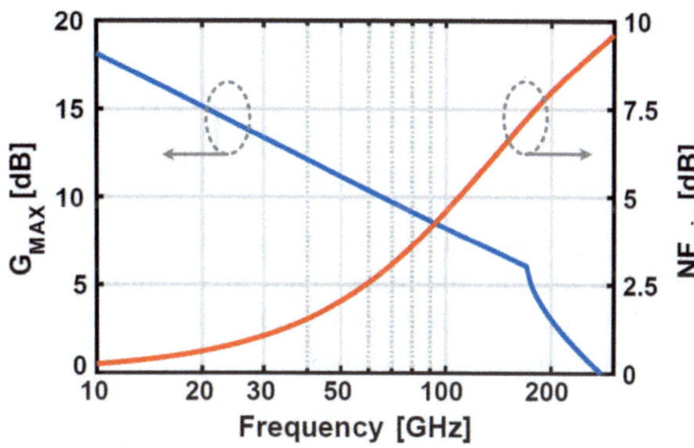

Fig. 1.5 G_{MAX} and NF_{min} of a single transistor (W/L = 1.05 × 24 μm/28 nm) common source amplifier against frequency. ©2017 John Wiley and Sons. Reprinted, with permission, from [3]

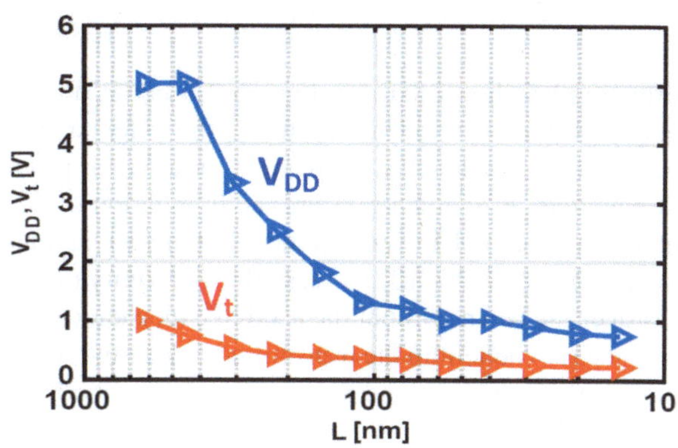

Fig. 1.6 V_{DD} and V_t against minimum channel length [10]

1.3 System Level Requirements for mm-Wave Wireless Links

1.3.1 Free Space Loss and Beamforming

Equation 1.2 shows that the path loss that the transmitted signal undergoes in free space increases with frequency. The attenuation is even more severe when the signal propagates through air (Fig. 1.2). It is instructive to focus on the following simplified example. Let us assume a line-of-sight communication link, where the transmitter (TX) and the receiver (RX) use antennas with directivity D_A to transmit a signal at a distance d. The Friis' equation [11] shows that the received power P_{RX} is

$$P_{RX} = \left(\frac{c\,D_A}{4\pi\,df}\right)^2 P_{TX}, \tag{1.4}$$

where P_{TX} is the transmitted power and f the frequency. For a given P_{TX}, the only way to overcome the higher path loss at higher frequencies without reducing the link distance, is to increase the antenna directivity. For a given antenna size A, the directivity can be expressed as

$$D_A = 4\pi\,A\left(\frac{f}{c}\right)^2. \tag{1.5}$$

Equation 1.5 shows that for a given A a better directivity is achieved at higher frequencies. Or for a given directivity, the antenna size gets smaller. This is a major benefit at system level, since in the same area it is possible to squeeze a larger number of antennas. Antenna arrays are the foundation of beamforming. By controlling the phase shift of the RF signal at the input of each antenna, it is possible to (1) combine the power of N TXs, (2) increase the directivity of both the TX and RX antenna and (3) steer the beam without the need of a mechanical actuator. An array of N elements provides N^2 benefit in the transmitted power and N times benefit in the RX power (at the RX the signal and the noise are both amplified, resulting a reduced benefit when compared to the TX). A N^3 benefit in the link budget is therefore expected, without sacrificing area and feature size, making mm-Wave communication links a real candidate for 5G.[1,2] It should be noted that CMOS state-of-the-art PAs shows lower

[1]In this example we assume that the integrated power amplifier is optimized for maximum linear output power and power added efficiency for a given technology and N PAs are integrated in the phased array.

[2]It is worth mentioning that an array with N elements used both at the TX and at the RX side does provide N^3 benefit in the link budget if and only if compared to a single antenna with an area A/N. Such a comparison is obviously not fair. When an antenna with the same area of the full array is used in combination with an ideal big PA that delivers N times larger output power, the N^3 benefit disappears. However, (1) phased arrays enable electrical beam steering and do not need a

output power when compared to power amplifiers implemented in other technologies (e.g. GaN, SiGe or GaAs). If on the one hand phased arrays permits the use of unit PAs with lower output power while still meeting the link budget requirements, on the other it should be noted that the DC power consumption also increases with N. As the power consumption increases so does the heat generation, seriously challenging the feasibility of massive MIMO system. Massive MIMO systems therefore need low power PA solution and deep-scaled CMOS technology is attracting an ever increasing attention.

1.3.2 Impairments Model

Equation 1.1 shows that the theoretical maximum channel capacity is proportional to the RF bandwidth of the signal. However, a practical modulation scheme can only get close to this maximum. The spectral efficiency measures how many bits can be squeezed in a given BW

$$BW = F_S\,(1+\alpha) = \frac{F_b}{log_2(M)}(1+\alpha), \tag{1.6}$$

where F_S is the symbol rate, α is the roll-off of the root raised cosine filter needed to limit the inter-symbol-interference (ISI) (typically 0.3$\sim$0.5) [9], F_b is the bit rate and M is the order of the M-QAM scheme adopted. Clearly, the higher the order M, the higher the spectral efficiency.

However, high order modulation schemes pose much higher requirements on each block of the system. To estimate the impact of noise and distortion over the bit error rate (BER) and derive system level requirements for each block, two main performance metrics are normally adopted. (1) Signal-to-Noise Ratio (SNR) and (2) Error Vector Magnitude (EVM). The latter is defined as the RMS magnitude of the error vector computed and expressed as a percentage of the EVM normalization reference (we will return on the intricacies of this definition later). Although SNR and EVM measure the same signal degradation, depending on the specific block considered (i.e. TX, RX or LO) it is preferable to refer to one of the two. In even-order M-QAM modulations, the bit-error rate (BER) can be approximated as follow [12]

$$BER \approx \frac{1}{log_2(M)}4\left(1-\frac{1}{\sqrt{M}}\right)Q\left(\sqrt{\frac{3}{M-1}SNR}\right), \tag{1.7}$$

---

mechanical actuator. Therefore, highly directive communication between the base station and the user equipment would be possible, enabling spacial reused. This technique in combination with the classical frequency and time reuse is expected to significantly increase the capacity of the whole wireless system. And, (2) as it will be discussed in Chap. 7, implementing a big PA that delivers N times larger output power at mm-Wave might not be possible or may result in unacceptably low efficiency.

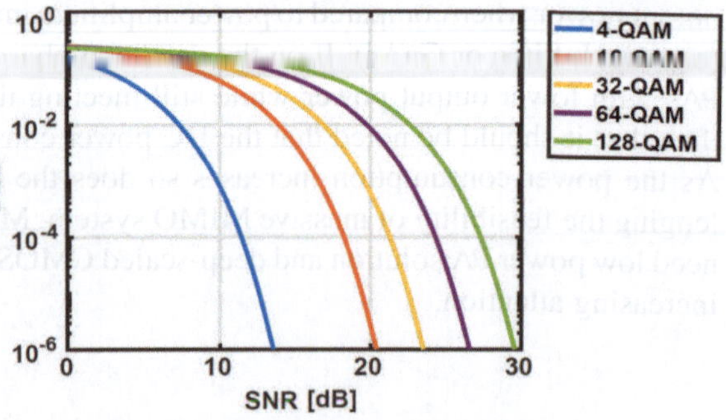

Fig. 1.7 BER versus SNR for different modulation schemes

where Q is the Q-function. Figure 1.7 shows the SNR requirements for different M-QAM schemes according to Eq. 1.7. Even when an ideal transceiver is adopted, a much higher SNR is needed to achieve the same BER as M increases.

Figure 1.8 shows a simplified block diagram of a wireless link, with a direct conversion TX and RX and a fundamental quadrature PLL. A line of sight communication with highly directive antennas (i.e. no multipath fading) will be considered in the following. Therefore, the channel adds white Gaussian noise only. The transmitter is the major responsible for distortion, maximum link distance (through PA output power), battery life time (PA efficiency) and I/Q amplitude and phase imbalance. The receiver is the major responsible for sensitivity (dominated by the LNA noise figure), link distance (through RX sensitivity), battery life time (LNA efficiency) and I/Q amplitude and phase imbalance. The fundamental quadrature PLL is the major responsible for phase noise (both at the TX and RX), battery life time (mm-Wave QVCO and divider power consumption), I/Q amplitude and phase imbalance. The digital baseband processing can partially compensate for PA non linearity (through pre-distortion techniques), QPLL phase noise at low frequency offset (through carrier tracking) and I/Q amplitude and phase imbalance.

The EVM of the whole system can be expressed as

$$EVM_{system} = \sqrt{\frac{1}{SNR_{system}}} = \\ = \sqrt{EVM^2_{AWGN} + EVM^2_{IQ} + EVM^2_{PN} + EVM^2_{PA}}. \tag{1.8}$$

The effect of each of these impairments on the signal integrity is the focus of this section.

1.3.2.1 Additive White Gaussian Noise (AWGN)

In absence of multipath fading, the channel can be model as an AWGN one. The effect of AWGN on the constellation is shown in Fig. 1.9a. The higher the noise,

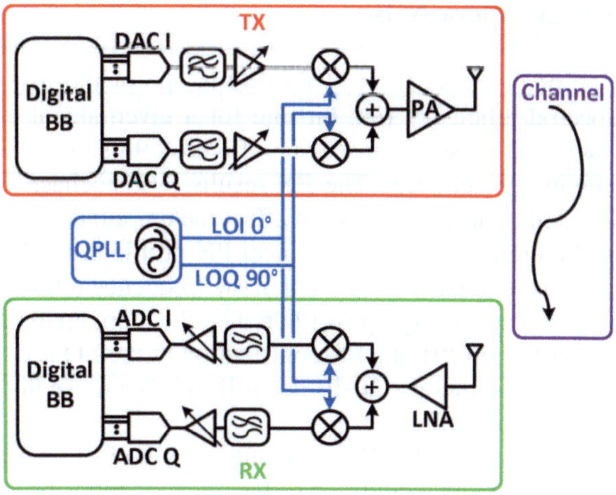

Fig. 1.8 Simplified wireless link block diagram

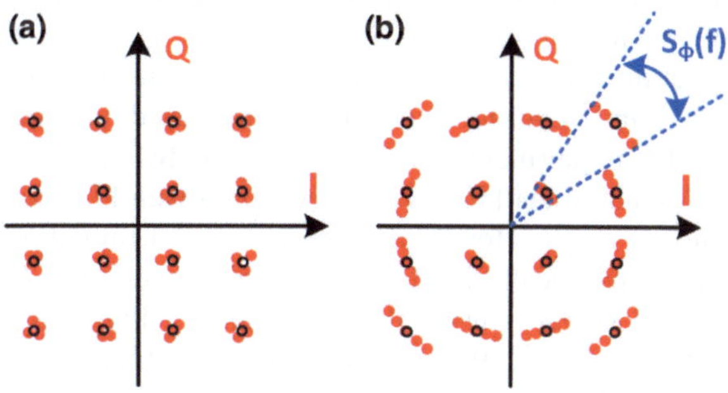

Fig. 1.9 **a** Effect of AWGN and **b** phase noise on the output constellation

the higher the BER as clear in Fig. 1.7. The noise of the system defines the receiver sensitivity, setting a fundamental limit to the link distance.

In presence of AWGN the link between EVM and SNR is simply [13]

$$EVM_{AWGN} = \sqrt{\frac{1}{SNR_{AWGN}}}. \tag{1.9}$$

1.3.2.2 I/Q Imbalance

The I/Q amplitude and phase imbalance result in a constant offset in the constellation point in amplitude and phase respectively. It is worth noting that the I/Q imbalance is the only impairment considered that is not stochastic or dependent on the modulation scheme adopted. Therefore, it is the easier to compensate for in the digital baseband.

1.3.2.3 Phase Noise

The phase noise (PN) of the LO is one of the major limitation to the maximum spectral efficiency (i.e. bit rate for a given signal bandwidth) obtainable in fully integrated CMOS transceivers. Figure 1.9b shows the effect of phase noise on the output constellation. The PN results in a stochastic rotation of the symbol in the constellation while does not affect the amplitude.

Figure 1.10 shows the typical PN profile at the PLL output. The close-in phase noise is dominated by the PN of the frequency reference used in the PLL. The PN of the oscillator is high-pass filtered by the loop, up to the PLL bandwidth [9]. Outside the PLL loop BW the PN contribution of the LO is dominant and in a well designed oscillator shows a −20 dB/dec roll-off. The noise floor is dominated by the thermal noise of the buffer.

To minimize the PN at the output of the PLL, a low-noise LO is needed. A wide loop bandwidth would also be beneficial to relax the requirements on the noise of the VCO. In state-of-the-art mm-Wave PLL, BW_{PLL} is normally limited to a maximum of 1∼3 MHz [14].

Intuitively, the low frequency PN[3] results in a slow movement of the symbols in Fig. 1.9b. A decision driven PLL can be used for symbol-timing recovery, mitigating the PN impairments up to a certain bandwidth BW_{TL}, drastically relaxing the PLL PN requirement [12, 13]. A decision driven PLL behaves as a 2nd order high-pass filter at BW_{TL}. However, there are several limitations to the maximum BW_{TL} that can be practically used. (1) For proper operation, $BW_{TL} << BW_{RF}$. (2) When an OFDM signal is used, the maximum tracking loop bandwidth is further reduced $BW_{TL} < BW_{sub}$. Where BW_{sub} may be estimated as half of the subcarrier spacing [13]. And (3), since the tracking loop behaves as a second PLL that uses the received data as reference, the phase component of the AWGN of the received signal is low-pass filtered and converted into phase noise [12] (see Fig. 1.10b). Therefore, there is a limit to the PN suppression that can be achieved with this technique. Moreover, for a given PLL phase noise profile, BW_{RF} and target SNR_{system} there is an optimal BW_{TL} that maximizes the PN suppression.

By referring to Fig. 1.10 and assuming a 2nd order tracking loop, the resulting EVM can be expressed as

$$EVM_{PN} = \sqrt{\frac{1}{SNR_{PN}}} = \sqrt{\frac{1}{SNR_{PLL}} + \frac{1}{SNR_{TL}}} = = \sqrt{2\int_{0}^{BW_{RF}/2} PN(f)df + 2\int_{0}^{BW_{RF}/2} \overline{V_n^2}(f)df}, \tag{1.10}$$

[3] Low frequency with respect to the modulation bandwidth of the signal BW_{RF}.

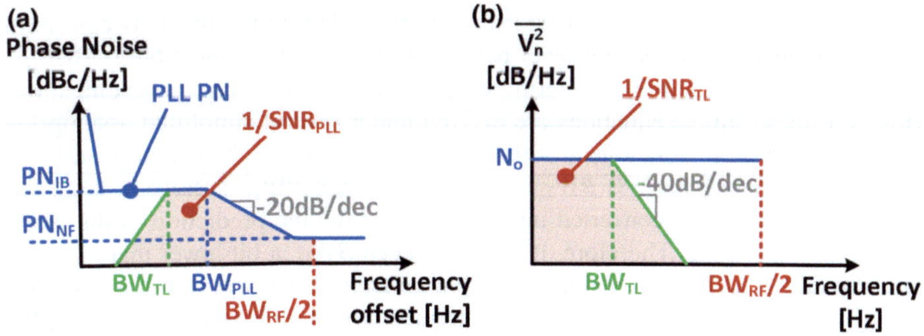

Fig. 1.10 **a** Typical phase noise profile at the PLL output when a data-aided 2nd order tracking loop in the digital baseband is applied. **b** Noise contribution of the carrier tracking loop in the digital baseband

where

$$\begin{aligned}\frac{1}{SNR_{PLL}} &= 2\int_{0}^{BW_{TL}} \frac{10^{PN_{IB}/10}}{BW_{TL}^4} f^4 df + 2\int_{BW_{TL}}^{BW_{PLL}} 10^{PN_{IB}/10} df + \\ &+2\int_{BW_{PLL}}^{f_{NF}} \frac{10^{PN_{IB}/10}}{f^2} BW_{PLL}^2 df + 2\int_{f_{NF}}^{BW_{RF}/2} 10^{PN_{NF}/10} df = \\ &= 2\frac{10^{PN_{IB}/10}}{5} BW_{TL} + 2\cdot 10^{PN_{IB}/10}(BW_{PLL} - BW_{TL}) + \\ &+2\cdot 10^{PN_{IB}/10} BW_{PLL}^2 \left(\frac{1}{BW_{PLL}} - \frac{1}{f_{NF}}\right) + \\ &+2\cdot 10^{PN_{NF}/10}\left(\frac{BW_{RF}}{2} - f_{NF}\right),\end{aligned} \tag{1.11}$$

$$f_{NF} = BW_{PLL}\sqrt{\frac{10^{PN_{IB}/10}}{10^{PN_{NF}/10}}}, \tag{1.12}$$

$$\begin{aligned}\frac{1}{SNR_{TL}} &= 2\int_{0}^{BW_{TL}} 10^{N_o/10} df + 2\int_{BW_{TL}}^{BW_{RF}/2} 10^{N_o/10}\frac{BW_{TL}^4}{f^4} df = \\ &= 2\cdot 10^{N_o/10} BW_{TL} + 2\cdot 10^{N_o/10}\frac{BW_{TL}^4}{3}\left(\frac{1}{BW_{TL}^3} - \frac{2^3}{BW_{RF}^3}\right),\end{aligned} \tag{1.13}$$

$$N_o = 10log_{10}\left(\frac{1}{2\,BW_{RF} SNR_{system}}\right). \tag{1.14}$$

The closed form expressions derived in Eqs. 1.10–1.14 provide a link between circuit design parameters and system level performance (i.e. EVM and SNR). They are therefore extremely helpful to determine a first estimate of the design specifications. However, these simple equations are derived under several simplified assumption. (1) The flicker noise component of the LO PN is neglected. (2) The in-band PN of the PLL is assumed flat. (3) $BW_{TL} << BW_{RF}$. When BW_{TL} is large, a large part of the channel noise is converted into phase noise and the prediction of the model may result inaccurate. Therefore, it is best to keep BW_{TL} a bit lower that the value that maximizes SNR_{PN}, so that the contribution from SNR_{PLL} is still dominant. For more accurate predictions a detailed system model in Matlab or Simulink should be adopted.

1.3.2.4 Distortion

The main contributor to the distortion of the whole system is the power amplifier at the transmitter side. Figure 1.11 shows the effect of the PA non-linearity on the signal constellation. The deviation of the constellation points from their ideal position is measured as EVM. Therefore, PAs typically use EVM as main performance parameter under modulated signal measurements. It is worth noting that the effect of distortion on the constellation points can be decomposed in two parts. (1) Effect on the amplitude, referred as AM-AM. And (2) effect on the phase, referred as AM-PM.

The effect on the output spectrum is shown in Fig. 1.12. Clearly, distortion raises the noise floor both in the channel of interest and in the adjacent channel, degrading the SNR. The major contribution to distortion in this case can be decomposed in 3rd and 5th order non linearity components [15].

The PA linearity together with the LO PN constitute the major bottlenecks of the system. Predistortion techniques can be applied in the baseband signal processing to

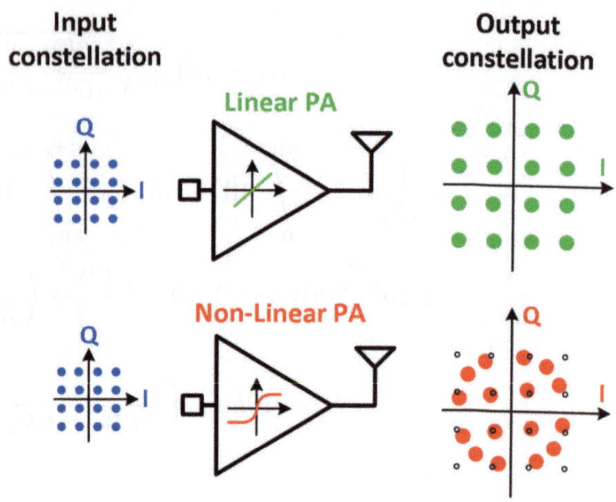

Fig. 1.11 Effect of PA distortion on the output constellation

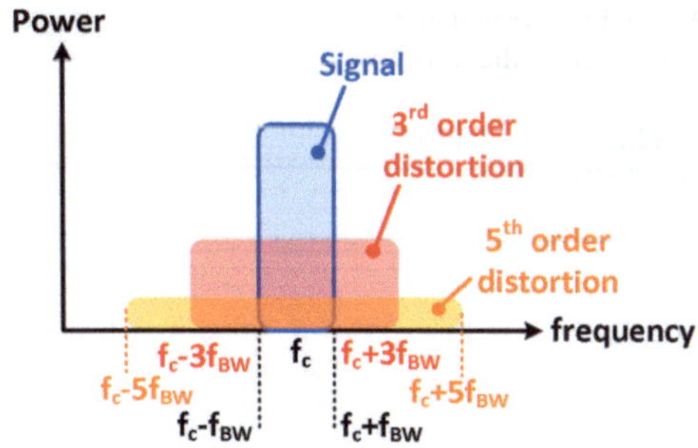

Fig. 1.12 Effect of PA distortion on the output spectrum

partially compensate for it. However, those techniques face several limitations. (1) They rely on complex signal processing, posing a limit to the effectiveness of the practical implementation. (2) The physical mechanism that cause distortion depend on temperature and vary during the life time of the PA. Therefore, the predistortion algorithms should be able to track these time-variant effects. And (3) those techniques should provide a compensation over the complete RF bandwidth of the modulated signal.

For all these reasons, predistortion alone can not possibly be the solution. In practical system, a back-off from the saturated power is applied, limiting further the trade-offs between average output power (needed in the link budget) and power added efficiency (PAE) (i.e. battery life-time). Further, modulation schemes with higher spectral efficiency shows larger peak-to-average power ratio (PAPR), requiring more margin from the saturated output power.

1.3.2.5 More on EVM Definitions

From the discussion above it is clear that the error vector magnitude (EVM) is a key indicator of modulated signal quality. It measures how far a transmitted or received constellation point is from its ideal location. Compared to other system-level specifications such as bit error rate, the EVM contains more information about amplitude and phase distortion and circuit limitations. It is designed to be a measurement of in-band signal quality. This is one of the major reasons why EVM is widely used to quantify the degradation of modulated signals due to circuit impairments, especially for transmitters and including the effects from power amplifiers (PAs). However, there are multiple ways to calculate EVM, and these methods do not provide identical results. Therefore it is important to be aware of these differences when a comparison with the state-of-the-art is made. For any performance comparison to be valid, it is essential to apply the exact same metric. Otherwise the comparison is not valid.

Within a specific communication standard the method to measure or calculate EVM is clearly indicated, but when no standard is available one must be careful. This is especially true for 5G the 5th generation of wireless systems which as of this moment of writing does not yet have a standard.

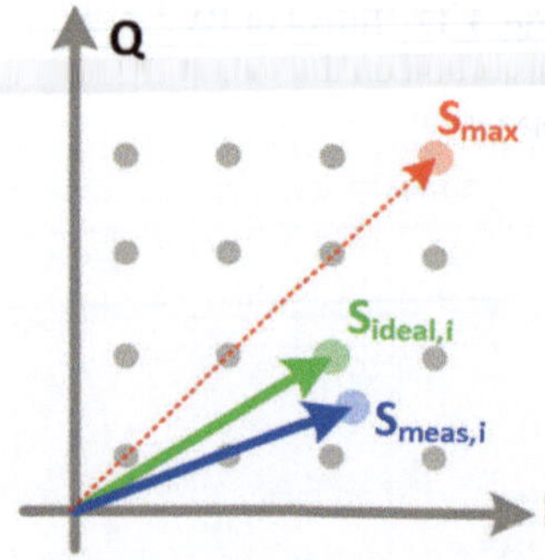

Fig. 1.13 Normalized constellation diagram for 64-QAM. Only the 1st quadrant is shown for simplicity

To get a better understanding of the different ways to calculate EVM, it is useful to briefly go back to the definition of EVM. Figure 1.13 shows the normalized 1st quadrant constellation diagram for a 64-QAM signal. At the optimal sampling point, one calculates the error vector as the difference between the measured and ideal symbol. This error vector is represented by a complex number based at the ideal constellation point. The EVM is now defined as a ratio of the RMS value of all the error vectors, averaged over N symbols, and then divided by some normalization factor (see Eqs. 1.15, 1.16). The calculation is done over many symbols to avoid the influence of bit pattern segments. The number of symbols N needs to be large enough to ensure that all possible symbols and transitions are observed.

Two EVM definitions are commonly used [16, 17]. The first one is a ratio of RMS magnitudes

$$EVM_{RMS} = \frac{\sqrt{\frac{1}{N}\sum_{i=1}^{N}|S_{ideal,i} - S_{meas,i}|^2}}{\sqrt{\frac{1}{N}\sum_{i=1}^{N}|S_{ideal,i}|^2}} = \frac{V_{error,RMS}}{C_{RMS}}, \tag{1.15}$$

where C_{RMS} is the RMS value of the constellation point magnitudes. The second compares the RMS magnitude of the errors to the peak magnitude of the constellation

$$EVM_{max} = \frac{\sqrt{\frac{1}{N}\sum_{i=1}^{N}|S_{ideal,i} - S_{meas,i}|^2}}{|S_{max}|} = \frac{V_{error,RMS}}{C_{max}}, \tag{1.16}$$

where $S_{ideal,i}$, $S_{meas,i}$ and S_{max} are defined for the ith symbol in Fig. 1.13.

EVM_{RMS} normalizes the RMS value of the error vectors to the RMS level of the M-ary signal constellation, while EVM_{max} adopts the maximum constellation magnitude as its normalization factor [16]. The two definitions coincide for constellations with constant magnitude (e.g. QPSK, BPSK, 8PSK, etc.), while $EVM_{RMS} > EVM_{max}$ for constellations with multiple possible magnitudes (e.g. APSK, Star-QAM, 16-QAM, 32-QAM, etc.). There also is a third EVM metric to add to this confusion: EVM_{peak} is the maximum value of the error vector magnitude that has occurred over

Table 1.1 Overview of constellation PAPR and signal PAPR numbers for 5 different constellation diagrams

Constellation diagram	PAPR of the ideal constellation (dB)	PAPR of RF signal envelope after SRRC ($\alpha = 0.35$) (dB)
QPSK	0	4
8PSK	0	4
16-QAM	2.6	6.6
64-QAM	3.7	7.7
256-QAM	4.2	8.2

sets of N symbols each. One must be particularly careful not to confuse EVM_{max} with EVM_{peak}.

From Eqs. 1.15 and 1.16 it is evident that the difference between EVM_{RMS} and EVM_{max} has something to do with the PAPR of the signal. This gives us the opportunity to address another point of confusion between electrical engineers having different backgrounds. The difference between EVM_{RMS} and EVM_{max} is not equal to the PAPR of the RF signal. Indeed, the difference between the two is equal to the PAPR of the ideal constellation diagram, i.e. before any Nyquist or channel filtering takes place. The PAPR of the constellation diagram itself can easily be calculated [18] and some numbers for well-known modulation formats are shown in Table 1.1. When the baseband filtering is applied, the PAPR of the signal increases above the PAPR of the constellation itself. For the examples shown in Table 1.1, this PAPR increase is 4 dB for a typical Square-Root-Raised Cosine filter with α equal to 0.35. It is worth noting that there is a key difference in the definition for PAPR for analog and RF designers. The PAPR for analog is equal to the square of the peak instantaneous voltage divided by the square of the RMS voltage value of the signal. But for RF designers the PAPR of a modulated carrier is defined differently. It is equal to the peak-envelope power (PEP) divided by the RMS power of the signal. PEP is the average power of a sinewave having an amplitude equal to the peak instantaneous voltage of the modulated carrier. Therefore, from an RF perspective, an unmodulated carrier has a PAPR of 1 (or 0 dB), whereas that same signal has a PAPR of 1.4 (or 3 dB) for an analog designer. This is to be expected, since a baseband OPAMP needs excellent circuit linearity to properly amplify a sinusoidal signal with constant envelope, while a bandpass RF PA does not require any circuit linearity to achieve the very same goal (see switch-mode power amplifiers [19]).

With the background developed so far, it is easy to derive $EVM_{RMS} \approx EVM_{max} +$ 2.6 dB for a 16-QAM, $EVM_{RMS} \approx EVM_{max} + 3.7$ dB for a 64-QAM and $EVM_{RMS} \approx EVM_{max} + 4.2$ dB for a 256-QAM. The EVM_{RMS} allows to better compare the signal quality for different modulation schemes and in presence on AWGN only is equal to -SNR [17].

A good example of confusion arising from using different EVM metrics is shown in Fig. 1.14. It shows three different measured constellation plots and the reported

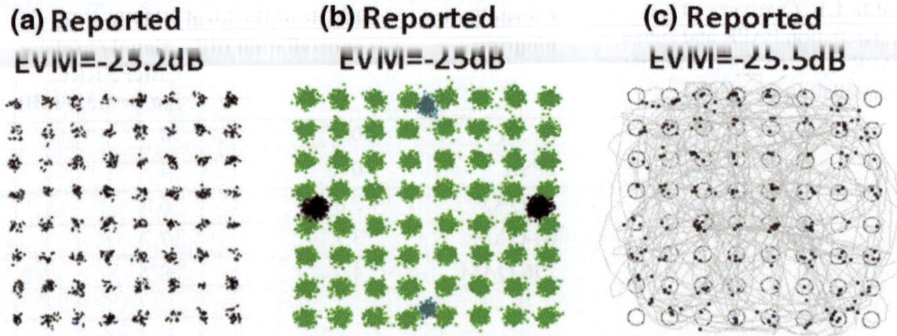

Fig. 1.14 Measured constellation and reported EVM of three state-of-the-art mm-Wave power amplifiers for future 5G presented at ISSCC in 2014 [20] (**a**) and 2016 [21] (**b**), [22] (**c**). Note that [22] **c** has no points at the constellation corners, clearly showing circuit compression, which is not seen in the other measurements (yet the same EVM value is reported)

EVM of three state-of-the-art mm-Wave power amplifiers developed for future 5G communications [20–22]. All of them amplify a 64-QAM modulated signal and provide measured results. All of them use a published definition of EVM and appear to achieve similar benchmark EVM numbers, all are around −25 dB. But clearly, for the same reported −25 dB EVM, the constellation plots look quite different. Indeed, the definition of EVM used in [20, 21] is EVM_{RMS}, whereas the one used in [22] is EVM_{max}. It is worth noting that any power amplifier design entails a stringent trade-off between efficiency and linearity. The requirements on output signal accuracy are often set by an EVM metric. To put things in perspective, −25 dB EVM allows 3 dB margin on the required SNR for a 64-QAM signal [21] when using EVM_{RMS}. This margin disappears if the EVM_{max} metric is used. To meet the specifications, a substantial power back-off from the maximum achievable output power is needed, compromising efficiency. Therefore, a ≈3.7 dB difference in the EVM definition immediately results in a very misleading comparison table, especially if the normalization used is omitted. Unfortunately, several comparison tables of this kind can be found to date in literature. Therefore, while we wait for the 5G standard to be released, it is important to clearly indicate the equation that is used to calculate EVM.

1.3.2.6 More on PAPRs

The PAPR of the signal plays a crucial role in defining the linearity requirements of the transmitter and of the PA in particular. Moreover, the PAPR of the signal sets the difference between the two discussed normalizations of EVM (i.e. EVM_{RMS} vs. EVM_{max}). However, it should be noted that the PAPR in the aforementioned cases refers to two different signals and results in general (and also in practice) in different values. To get more insight, Fig. 1.15 shows the simplified block diagram of a direct-conversion transmitter for mm-Wave applications emphasizing different

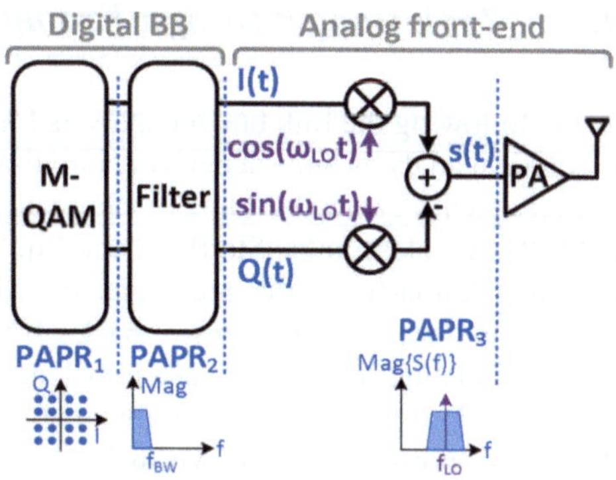

Fig. 1.15 Simplified block diagram of an I/Q mm-Wave direct-conversion transmitter

signals present at different sections. The signal at the PA input can be written as [18]

$$\begin{aligned} s(t) &= I(t)cos(\omega_{LO}t) - Q(t)sin(\omega_{LO}t) = \\ &= Re[I(t) + Q(t)]e^{j\omega_{LO}t} = r(t)cos(\omega_{LO}t + \theta(t)), \end{aligned} \tag{1.17}$$

where $r(t) = \sqrt{I^2(t) + Q^2(t)}$ is the envelope of the baseband signal. It is possible to show that when the bandwidth of the BB signal is $f_{BW} \ll f_{LO}$ the PAPR of the RF signal at the PA input can be written as [23]

$$PAPR(s) = 10log_{10}\left(\frac{2P_{peak}(r)}{P_{RMS}(r)}\right) = PAPR(r) + 3\,\text{dB}. \tag{1.18}$$

The difference between EVM_{RMS} versus EVM_{max} is equal to the PAPR of the ideal constellation diagram, therefore the baseband signal with $PAPR_1$ depicted in Fig. 1.15 should be considered. Before being upconverted to RF, this signal is low-pass filtered to limit its bandwidth [9, 18]. Thus, the PAPR of the baseband signal envelope $PAPR_2$ (in Fig. 1.15) is equal to $PAPR(r)$ in Eq. 1.18, and typically remarkably higher than $PAPR_1$ [18] as reported in Table 1.1. Finally, when the baseband definition of PAPR is used and under the assumption of $f_{BW} \ll f_{LO}$, the upconverted signal shows a $PAPR_3$ ≈3 dB higher than $PAPR_2$.

As it will be shown later in the chapter dedicated to power amplifiers, even if the signal at the PA input shows the PAPR derived in Eq. 1.18, the real challenge is to amplify a signal with a non-constant envelope. When a PA is modeled as an hard limiter, to guarantee an ideally linear amplification, the back-off needed form the saturation point is indeed equal to the PAPR of the envelope of the baseband signal (and not 3 dB higher than that) [24]. This is the reason why a different definition of PAPR is used for RF band-pass signals. To an RF designer $PAPR_2 = PAPR_3$.

1.3.3 Link Budget Design Examples

In the following the link budget analysis for two mm-Wave links is carried out. This theoretical analysis aims at deriving circuit level specification for the most important high frequency analog building blocks in the PLL, TX and RX for both an E-Band and a 32 GHz wireless links. Starting from Eq. 1.8, the following assumptions are made. (1) The PA non-linearities are neglected. Meaning that predistortion techniques are applied and/or a sufficient power back-off from the saturated output power is taken. (2) The I/Q amplitude and phase imbalance is compensated by the baseband digital circuitry. (3) A line-of-sight communication is considered in an AWGN channel. (4) The phase noise profile shown in Fig. 1.10 is considered, with $BW_{PLL} = 1\,\text{MHz}$, $BW_{TL} = 300$ kHz and $PN_{NF} = -150\,\text{dBc/Hz}$.

SNR_{system} in Eq. 1.8 in this case is composed by two contributors SNR_{AWGN} and SNR_{PN}. The higher the LO PN, the higher SNR_{AWGN} needs to be to guarantee the required BER. The SNR degradation due to phase noise can be expressed in dB as [12]

$$10log_{10}\left(\frac{SNR_{system}}{SNR_{AWGN}}\right) = 10log_{10}\left(1 - \frac{SNR_{system}}{SNR_{PN}}\right). \tag{1.19}$$

Figure 1.16 shows the SNR degradation due to phase noise at 10^{-3} and 10^{-6} BER for different modulation schemes.[4,5] Clearly, the impact of PN is not negligible. As it will be shown in Chap. 4, $-110\,\text{dBc/Hz}$ at 10 MHz offset is a tough specification for a mm-Wave integrated oscillator, especially when a large tuning range is needed.

The estimated SNR required to achieve 10^{-3} and 10^{-6} BER is summarized in Table 1.2. Three cases are considered, (1) no PN, (2) $-110\,\text{dBc/Hz}$ and (3) $-120\,\text{dBc/Hz}$ at 10 MHz offset. An oscillator with $-110\,\text{dBc/Hz}$ PN at 10 MHz offset from the carrier can guarantee a 64-QAM communication at 10^{-3} BER and a 16-QAM at 10^{-6} BER. An oscillator with $-120\,\text{dBc/Hz}$ PN at 10 MHz offset from the carrier can guarantee a 128-QAM communication at 10^{-6} BER. In both cases, the noise floor far away from the carrier has a negligible impact.

Once the minimum required SNR to meet the BER specification is estimated, the receiver sensitivity can be derived as follow

$$RX_{Sensitivity} = 10log_{10}(K_B T) + 30 + 10log_{10}(BW) + NF_{RX} + SNR_{min}, \tag{1.20}$$

where K_B is the Boltzmanns constant, T is the absolute temperature, BW is the RF bandwidth of the signal and NF_{RX} is the noise figure of the full receiver. It is useful to refer to the simplified schematic shown in Fig. 1.17 to finalize the design examples.

---

[4] In this design example we consider a modulated signal bandwidth of 500 MHz. However, when the BW_{RF} is increased to 4.75 GHz, the SNR degradation due to PN does not change significantly.

[5] It is worth mentioning that in this study we consider the effect of the PN of a single PLL on the SNR of the full system. When two PLLs with the same phase noise profile are used for the TX and RX paths, 3 dB better PN is needed to keep the same SNR.

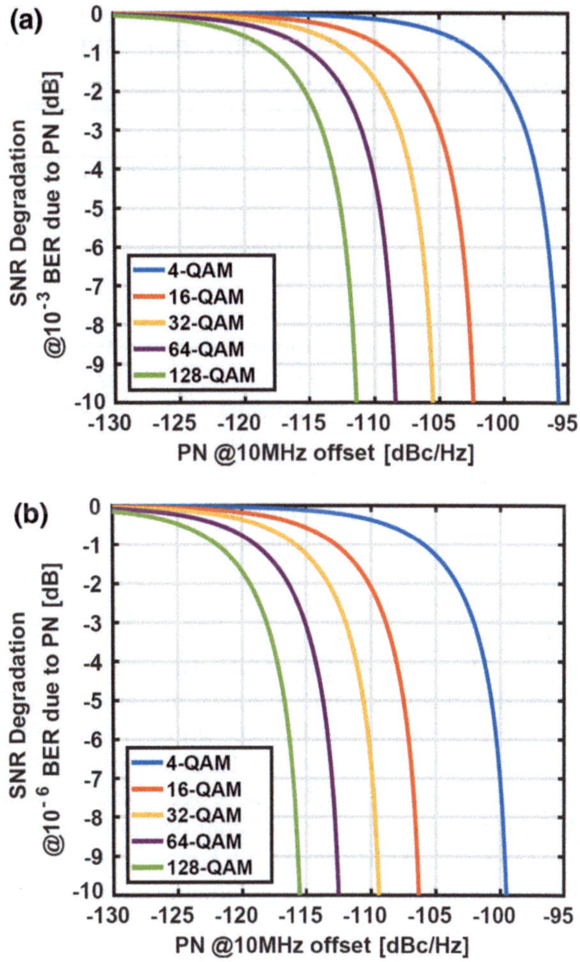

Fig. 1.16 SNR degradation due to phase noise at **a** 10^{-3} and **b** 10^{-6} BER

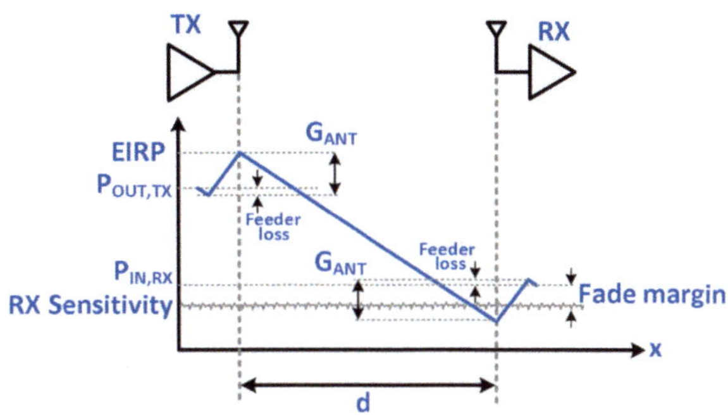

Fig. 1.17 Typical link budget design example

Table 1.2 Effect of different phase noise profiles on required SNR and BER

Modulation	PN @10 MHz offset	SNR_{min} @10^{-3} BER	SNR_{min} @10^{-6} BER
4-QAM	NO PN	9.8	13.6
	−120 dBc/Hz	9.8	13.6
	−110 dBc/Hz	9.9	14
16-QAM	NO PN	16.5	20.4
	−120 dBc/Hz	16.6	20.6
	−110 dBc/Hz	17.2	22.5
32-QAM	NO PN	19.6	23.5
	−120 dBc/Hz	19.7	23.8
	−110 dBc/Hz	21.3	30.2
64-QAM	NO PN	22.5	26.6
	−120 dBc/Hz	22.8	27.4
	−110 dBc/Hz	26.8	n.a.
128-QAM	NO PN	25.5	29.4
	−120 dBc/Hz	26.1	31.1
	−110 dBc/Hz	n.a.	n.a.

The 71 to 76 and 81 to 86 GHz frequency bands are divided in the US into four 1.25 GHz channels (eight in total). In Europe, a 125 MHz guard band is required at both ends of the 5 GHz spectrum to prevent potential interference to and from adjacent bands. The two 4.75 GHz bands are further divided into nineteen 250 MHz channels. All the channels may be aggregated without limit both the US and Europe [6]. In the following, we will consider an E-Band wireless link that employs two channels of 4.75 GHz bandwidth between 71 to 76 GHz and 81 to 86 GHz. The link distance is set to $d = 1$ km, resulting in a free space path loss of

$$FSPL = 92.4 + 20log_{10}(d_{km}) + 20log_{10}(f_{c,GHz}) = 131\,\text{dB}. \qquad (1.21)$$

The atmospheric attenuation is 0.3 dB/km and the rain attenuation considered to guarantee a for 99.999% weather availability in London (5 min of outage per year) is 21.4 dB/km [25]. The output referred 1 dB compression point of the PA is set to 20 dBm, the receiver NF is 10 dB and the off-chip antenna gain is 50 dBi. 5 dB feeder and implementation losses are considered at the RX and TX side respectively, resulting in 10 dB total loss. The channel noise is $10log_{10}(K_B\,T\,BW) + 30 = -77.2$ dBm. −110 dBc/Hz PN at 10 MHz offset is considered for the PLL. To satisfy the linearity requirements for the PA, a back-off from P_{out} equal to the signal PAPR is assumed.

Table 1.3 shows the predicted bit rate and fade margin with and without rain for different modulation schemes. This system could provide a 1 km wireless link featuring up to 35 Gb/s @10^{-6} BER using a 32-QAM under good weather condition and 14 Gb/s @10^{-6} BER using a 4-QAM under heavy rain.

Table 1.3 E-Band link budget. $BW_{RF} = 4.75\,GHz$, 2-channel bonding

Modulation	PAPR (dB)	Bit rate (Gb/s)	Fade margin @10^{-3} BER	Fade margin @10^{-6} BER
4-QAM (w/rain)	4	14	32.2 dB (10.8 dB)	28.1 dB (6.7 dB)
16-QAM (w/rain)	6.6	28	22.3 dB (0.9 dB)	17 dB (n.a.)
32-QAM (w/rain)	6.3	35	18.6 dB (n.a.)	9.6 dB (n.a.)
64-QAM (w/rain)	7.7	42	11.6 dB (n.a.)	n.a. (n.a.)

Table 1.4 32 GHz link budget. $BW_{RF} = 500\,MHz$, 16 elements array

Modulation	PAPR (dB)	Bit rate (Gb/s)	Fade margin @10^{-3} BER
4-QAM	4	0.74	20.5 dB
16-QAM	6.6	1.37	10.6 dB
32-QAM	6.3	1.85	6.8 dB
64-QAM	7.7	2.22	n.a.

Following a similar procedure, we now focus on a 32 GHz wireless link with a 16 elements phased array. The RF signal bandwidth is set to 500 MHz. The frequency band from 31.8 to 33.4 GHz has been selected as a high priority channel for future 5G communications links above 24 GHz by the European platform METIS project [26]. The following assumptions are made. 10 m line of sight link, an atmospheric attenuation of 0.06 dB/km, a PA with an output referred 1 dB compression point of 13 dBm, 10 dB receiver noise figure and 5 dBi off-chip single-element patch antenna gain. 2.5 dB feeder and implementation losses are considered at the RX and TX side respectively.

Table 1.4 shows the predicted bit rate and fade margin for different modulation schemes. This system could provide up to 1.85 Gb/s wireless link over 10 m distance @10^{-3} BER using a 32-QAM with a 16 elements phased array.

1.4 Outline of This Book

This manuscript focuses on design challenges and techniques for mm-Wave building blocks for future 5G transceivers implemented in deep-scaled CMOS technology. In this chapter we discussed several reasons why CMOS will be a key enabler of 5G, and highlighted several challenges that need to be addressed both at circuit level and at system level. The rest of this book is organized as follow.

Chapter 2 reviews the major implications that aggressive technology scaling has on active and passive devices. The inversion coefficient (IC) is adopted as a design parameter to achieve optimal analog performance while allowing a simple comparison with different technologies. Moreover, it is shown that technology scaling does not provide any obvious benefit for passive devices.

To achieve the 5G requirements of 100× higher data rate, 100× more connected devices, 100× higher network efficiency while ensuring <1 ms latency design techniques for low power broadband mm-Wave front-ends are required. Gain-bandwidth (GBW) enhancement techniques are the object of Chap. 3. A strong focus is put on state-of-the-art techniques that lead to low insertion loss practical on-chip implementation. Several 4th order filters are compared and second order effects relevant to mm-Wave designers are discussed in great detail. Furthermore, simple design techniques to realize broadband impedance transformation, power dividers and combiners are introduced. This chapter together with Chap. 2 forms the foundation of the prototypes shown in the following chapters.

The basics of integrated mm-Wave oscillators and state-of-the-art tuning extension techniques are briefly recalled in Chap. 4. This chapter will conclude with the discussion of design, layout and measurements of an E-Band fundamental quadrature VCO implemented in 28 nm bulk CMOS. This oscillator covers two bands separated in frequency, while achieving low phase noise and accurate quadrature phases. When the silicon area is considered, this work achieves a measured FOM_A over the tuning range between 3.6 and 12.8 dB higher than the best previously reported one.

mm-Wave dividers are needed to close the loop of any fundamental phased locked loop for mm-Wave applications. The basics of high speed dividers are the focus of Chap. 5. The implications of aggressive CMOS technology scaling on divider design are shown. A broadband tunable divide-by-4 circuit implemented in 28 nm bulk CMOS is discussed. This work introduces simple design guidelines to realize a compact inductor-less divider that covers the whole E-Band (60–90 GHz) with wide margin while achieving state-of-the-art power consumption.

Chapter 6 is dedicated to design techniques for broadband low-noise power amplifiers and downconverters. The design, layout and measurements of two 28 nm bulk CMOS prototypes that demonstrate the proposed concepts are discussed. The first test chip is an E-Band LNA that achieves a measured figure of merit ≈10.5 dB better than state-of-the-art designs in the same band and comparable to LNAs at lower frequencies. The second test chip demonstrates for the first time a single-chip broadband receiver suitable for E-Band point-to-point communication links in deep-scaled bulk CMOS.

The power amplifier is a key bottleneck for power consumption, distortion and achievable link distance in any transmitter. The basics of PA design, the major causes of AM-PM distortion and state-of-the-art linearization techniques are discussed in Chap. 7. This chapter will conclude with the design, layout and measurements details of a 29–57 GHz (65% BW) AM-PM compensated class-AB power amplifier tailored for 5G phased arrays. High output power, high in-band and out-of-band linearity under wideband modulated signal are demonstrated, despite the 0.9 V supply and being realized in a 28 nm bulk CMOS process without RF thick top metal.

Chapter 8 summarizes the major contributions of this work and proposes some ideas for future work.

References

1. S. Onoe, 1.3 evolution of 5G mobile technology toward 1 2020 and beyond, in *2016 IEEE International Solid-State Circuits Conference (ISSCC), San Francisco, CA* (2016), pp. 23–28
2. ITU-R Recommendation M.2083-0, IMT vision - framework and overall objectives of the future development of IMT for 2020 and beyond (2015), p. 21
3. P. Reynaert, W. Steyaert, M. Vigilante, *RF CMOS. Nanoelectronics: Materials, Devices, Applications, 2 Volumes* (2017)
4. W.M. Holt, 1.1 Moore's law: a path going forward, in *2016 IEEE International Solid-State Circuits Conference (ISSCC), San Francisco, CA* (2016), pp. 8–13
5. C.E. Shannon, A mathematical theory of communication. Bell Syst. Tech. J. **27**(3), 379–423 (1948)
6. J. Wells, *Multigigabit Microwave and Millimeter-Wave Wireless Communications* (Artech House, Boston, 2010)
7. L. Reger, 1.4 the road ahead for securely-connected cars, in *2016 IEEE International Solid-State Circuits Conference (ISSCC), San Francisco, CA* (2016), pp. 29–33
8. W. Sansen, 1.3 analog CMOS from 5 micrometer to 5 nanometer, in *2015 IEEE International Solid-State Circuits Conference - (ISSCC) Digest of Technical Papers, San Francisco, CA* (2015), pp. 1–6
9. B. Razavi, *RF Microelectronics*, 2nd edn. (Prentice Hall, New Jersey, 2011)
10. ITRS, International technology roadmap for semiconductors, http://www.itrs.net/reports.html
11. D.M. Pozar, *Microwave Engineering* (Wiley, New York, 2009)
12. L. Iotti, A. Mazzanti, F. Svelto, Insights into phase-noise scaling in switch-coupled multi-core LC VCOs for E-Band adaptive modulation links. IEEE J. Solid-State Circuits **52**(7), 1703–1718 (2017)
13. T. Siriburanon et al., A low-power low-noise mm-wave subsampling PLL using dual-step-mixing ILFD and tail-coupling quadrature injection-locked oscillator for IEEE 802.11ad. IEEE J. Solid-State Circuits **51**(5), 1246–1260 (2016)
14. W. Wu, R.B. Staszewski, J.R. Long, A 56.4-to-63.4 GHz multi-rate all-digital fractional-N PLL for FMCW radar applications in 65 nm CMOS. IEEE J. Solid-State Circuits **49**(5), 1081–1096 (2014)
15. S.C. Cripps, *Advanced Techniques in RF Power Amplifier Design* (Artech House, Boston, 2002)
16. Agilent, Vector signal analysis basics application note 150-15, http://cp.literature.agilent.com/litweb/pdf/5989-1121EN.pdf
17. M.D. McKinley et al., EVM calculation for broadband modulated signals, in *64th ARFTG Conference Digest 2004*
18. E. McCune, *Practical Digital Wireless Signals* (Cambridge University Press, Cambridge, 2010)
19. E. McCune, A technical foundation for RF CMOS power amplifiers: part 5: making a switch-mode power amplifier. IEEE Solid-State Circuits Mag. **8**(3), 57–62 (Summer 2016)
20. S. Kulkarni, P. Reynaert, 14.3 a push-pull mm-wave power amplifier with <0.8 AM-PM distortion in 40 nm CMOS, in *2014 IEEE International Solid-State Circuits Conference Digest of Technical Papers (ISSCC), San Francisco, CA* (2014), pp. 252–253
21. S. Shakib, H.C. Park, J. Dunworth, V. Aparin, K. Entesari, A highly efficient and linear power amplifier for 28-GHz 5G phased array radios in 28-nm CMOS. IEEE J. Solid-State Circuits **51**(12), 3020–3036 (2016)
22. C.R. Chappidi, K. Sengupta, 20.2 a frequency-reconfigurable mm-wave power amplifier with active-impedance synthesis in an asymmetrical non-isolated combiner, in *2016 IEEE International Solid-State Circuits Conference (ISSCC), San Francisco, CA* (2016), pp. 344–345

23. T.J. Rouphael, *RF and Digital Signal Processing for Software-Defined Radio: a Multi-standard Multi-mode Approach* (Newnes, Amsterdam, 2009)
24. P. Reynaert, M. Steyaert, *RF Power Amplifiers For Mobile Communications* (Springer Science & Business Media, New York, 2006)
25. ITU-R P.837-4, Characteristics of precipitation for propagation modelling (2003)
26. ICT-317669-METIS/D5.1, Intermediate description of the spectrum needs and usage principles (2013)

Chapter 2
G_m Stage and Passives in Deep-Scaled CMOS

CMOS technology scaling allows faster transistors at each node, making mm-Wave analog design possible. However, scaling does not provide only benefits. The lower break down voltage forces the scaling of the voltage power supply as well, posing severe limitations on linearity, device stacking and achievable signal-to-noise ratio.

The back end of line (BEOL) metal stack gets thinner and closer to the substrate, making the effect of interconnection losses and parasitics dominant. Moreover, mm-Wave design is to some extend an upside down world when compared to RF design (in the low GHz range). At RF frequencies, capacitors show higher quality factor when compared to inductors. On-chip transmission lines are almost impossible to realize due to the large wavelength. At mm-Wave however the scenario is completely the opposite. Therefore, new design techniques are needed to face such technology constrains.

This chapter deals with the basic blocks available to analog designers in deep-scaled CMOS. The active devices are the focus of Sect. 2.1. Passive devices are discussed in Sect. 2.2. The aim is to briefly recall the basic of operation with a strong focus on the major challenges that a designer faces at mm-Wave. The effect of scaling is also discussed, leading to simple design guidelines and establishing the foundation of the following chapters.

2.1 G_m Stage: MOS as a Transconductor

Before diving into the operation of the MOS as an analog amplifier, it is useful to briefly recall some technology parameter. Those parameters describes the physics of the transconductor and are beyond the reach of the designer. The thoughtful derivation is beyond the scope of this work and can be found in [1].

M. Vigilante and P. Reynaert, *5G and E-Band Communication Circuits in Deep-Scaled CMOS*, Analog Circuits and Signal Processing,
https://doi.org/10.1007/978-3-319-72646-5_2

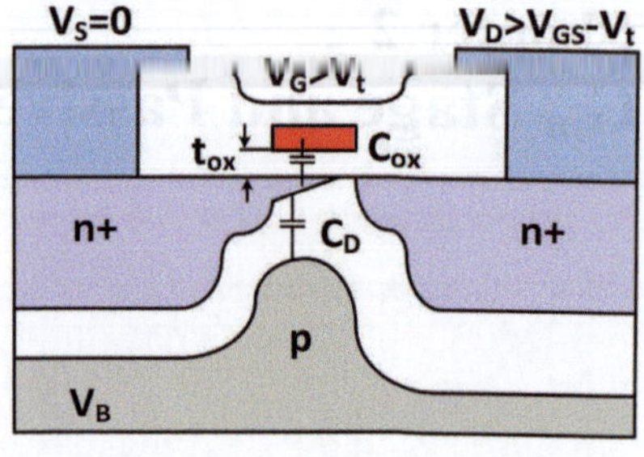

Fig. 2.1 Cross section of a NMOS transistor in saturation [1]

The cross section of a NMOS transistor in saturation (i.e. $V_{GS} - V_t > 0$ and $V_{DS} > V_{GS} - V_t$, where V_t is the threshold voltage) is shown in Fig. 2.1. It is possible to define the following technology parameters. $C_D = \varepsilon_{si}/t_{si}$, $C_{ox} = \varepsilon_{ox}/t_{ox}$, $n = C_D/C_{ox} + 1$, $K' = (\mu_n C_{ox})/(2n)$ and $U_T = (K_B\,T)/q$. Where ε_{si} and ε_{ox} are the silicon and oxide dielectric constants, t_{si} and t_{ox} are the depletion layer and oxide thicknesses, μ_n is the electron mobility, K_B is Boltzmanns constant, T is the absolute temperature and q is the electron charge.

2.1.1 DC Model and Regions of Operation (I_{DS})

Figure 2.2 shows the schematic of a single transistor common source (CS) amplifier and its DC I_{DS} versus V_{GS} plot. Clearly, for a given $V_{DS} > 0$, the output current I_{DS} increases with V_{GS}.

Four regions of operation are highlighted [1]. (1) The MOS is OFF when $I_{DS} = 0$. For increasing values of V_{GS} the transistor undergoes (2) weak-inversion (WI), (3) strong-inversion (SI) and (4) velocity saturation (VS). In each of these bias regions, the output current is found as

$$I_{DS,WI} = K'\frac{W}{L}(2n\,U_T)^2\,e^{V_{GS}/(n\,U_T)}, \tag{2.1}$$

$$I_{DS,SI} = K'\frac{W}{L}(V_{GS} - V_t)^2, \tag{2.2}$$

$$I_{DS,VS} = W\,C_{ox}\,v_{sat}(V_{GS} - V_t), \tag{2.3}$$

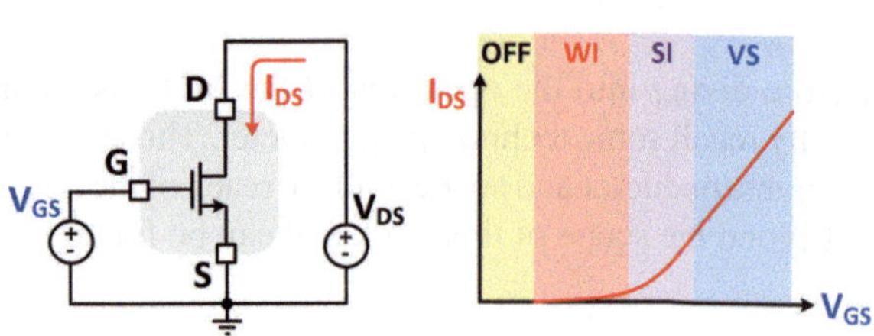

Fig. 2.2 NMOS common source (CS) amplifier, schematic and DC I_{DS} versus V_{GS} plot [1]

where W and L are the transistor width and length respectively. It is worth noting that even if the output current keeps increasing with V_{GS}, it first shows an exponential growth (in WI), then quadratic (in SI) and finally linear (in VS).

2.1.2 AC Model, Gain (g_m) and Speed (f_t, f_{MAX})

Figure 2.3 shows the simplified AC model of a CS amplifier. At low frequencies the capacitors behave as open circuits, and $V_{CGS} = V_{GS}$. The transconductance is found in each region of operation as $g_m = dI_{DS}/dV_{GS}$

$$g_{m,WI} = K'\frac{W}{L}4n\,U_T\,e^{V_{GS}/(n\,U_T)}, \tag{2.4}$$

$$g_{m,SI} = K'\frac{W}{L}2(V_{GS} - V_t), \tag{2.5}$$

$$g_{m,VS} = W\,C_{ox}\,v_{sat}. \tag{2.6}$$

These equations reveal insight into the operation of a single transistor MOS amplifier. The transconductance g_m is one of the most important design parameters and the price to pay is DC power consumption, I_{DS}.

Besides gain, another key parameter at mm-Wave is speed. The two most popular metrics of speed are f_t and f_{MAX}. The former is mainly technology dependent, whereas the latter contains more information about the layout parasitics and is partially under the control of the designer.

f_t is defined as the frequency for which the current gain is equal to 1. When the AC model in Fig. 2.3 is used and C_{GD} is neglected, $f_t = g_m/(2\pi C_{GS})$. By substituting $C_{GS} = (2/3)WLC_{ox}$ in Eqs. 2.4–2.6, it is possible to write

$$f_{t,WI} = \frac{3\mu}{2\pi L^2}e^{V_{GS}/(n\,U_T)}, \tag{2.7}$$

$$f_{t,SI} = \frac{3\mu}{4\pi L^2}(V_{GS} - V_t), \tag{2.8}$$

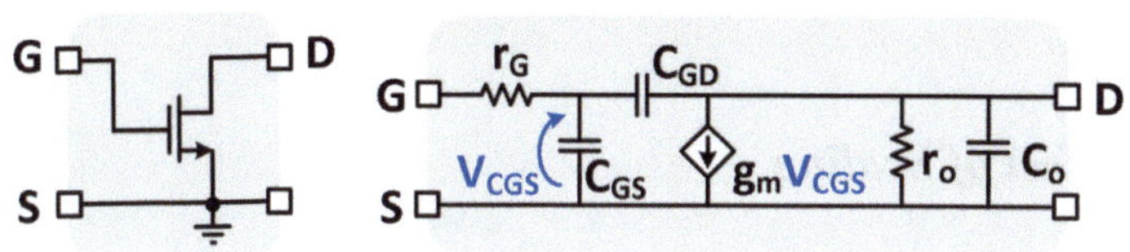

Fig. 2.3 Simplified AC model of a CS amplifier

$$f_{t\,VS} = \frac{v_{sat}}{2\pi L}. \qquad (2.9)$$

These equations show clearly the benefit of scaling the channel length L on speed. It is worth noting that in velocity saturation f_t is inversely proportional to L, whereas in WI and SI is inversely proportional to L^2.

f_{MAX} is defined as the frequency for which the maximum power gain is equal to 1. When the AC model in Fig. 2.3 is considered and C_{GD} is neglected,

$$f_{MAX} = \frac{f_t}{2}\sqrt{\frac{r_o}{r_G}}. \qquad (2.10)$$

Equation 2.10 shows that f_{MAX} depends on r_G, demonstrating the importance of the layout parasitics for high speed design. It is worth noting that $r_o = V_E\,L/I_D$ [1] and degrades with a smaller channel length.

2.1.3 Inversion Coefficient (IC) as a Design Parameter

To get more insight into the transistor operation and derive simple design guidelines, it is useful to normalize the output current against transistor width (W) and technology parameters [2]. This new design parameter is called Inversion Coefficient (IC) and it is defined as

$$IC = \frac{I_{DS}}{I_{spec}} = \frac{I_{DS}}{K'(W/L)\,(2n\,U_T)^2}. \qquad (2.11)$$

The bias point at which the MOS enters the velocity saturation region can be now expressed as

$$IC_{VS} = \frac{1}{\lambda_c^2} = \left(\frac{v_{sat}\,L}{2\mu\,U_T}\right)^2. \qquad (2.12)$$

The major performance parameters for a single transistor amplifier against IC are reported in Fig. 2.4 [3]. Gain and speed are precious at mm-Wave. Therefore, designers are willing to pay high DC power consumption to achieve the required performance. However, as soon as the transconductor enters the velocity saturation region, there is no benefit in increasing further the bias current.

2.1.4 Effect of Scaling

Technology scaling improves the speed of the transistor as clear from Eqs. 2.7–2.9. However, Eq. 2.12 shows that the smaller the channel length, the sooner the transistor

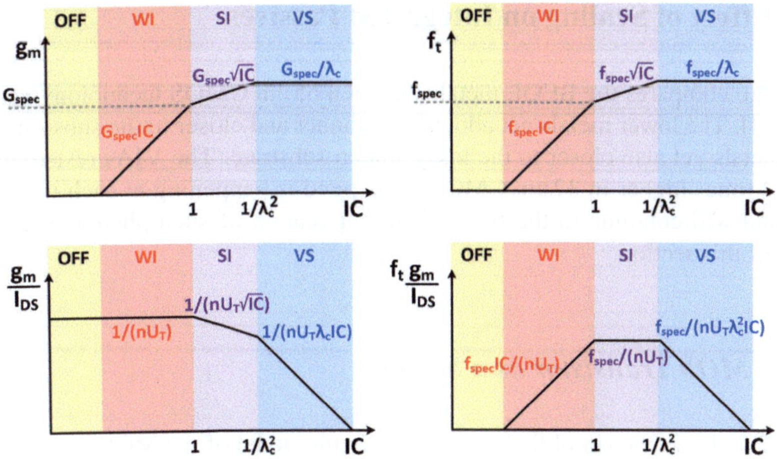

Fig. 2.4 Main single transistor amplifier performance parameters against IC. $G_{spec} = I_{spec}/(n\,U_T)$, $f_{spec} = \mu\,U_T/(\pi\,L^2)$ [3]

enters the velocity saturation region. Therefore, in deep-scaled CMOS the benefit in terms f_t is less evident than in the past. This phenomenon is graphically shown in Fig. 1.4.

One of the most powerful implication of adopting IC as a design parameter is that it is particularly simple to compare different technologies and predict how the analog design will evolve in the future. This trend is reported in Fig. 2.5 [3]. Clearly, in deep-scaled CMOS the strong inversion region is disappearing. Therefore, even mm-Wave analog amplifier are going to be designed deeper and deeper in weak inversion. This a key difference from the past.

It is worth noting that in 65 nm there is a flat optimum region that gives the best gain and speed performance for give power consumption. However, in 20 nm CMOS this is not the case anymore. Meaning that the designer should take extra care and choose the bias point that results in a rather sharp optimum.

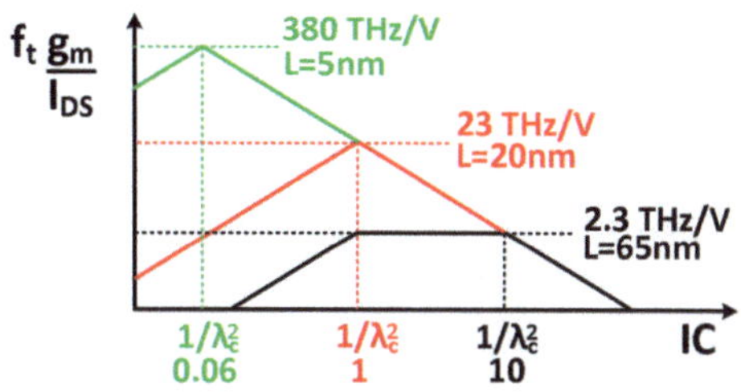

Fig. 2.5 $f_t\, g_m/I_{DS}$ against IC for different technology nodes [3]

2.2 Effect of Scaling on Integrated Passives

Figure 2.6 compares the BEOL metal stack of a 65 nm CMOS technology against a 32 nm [4]. The lower metals get about 50% thinner and closer to the substrate. High level metals get also closer to the lossy silicon substrate. The VIAs resistivity gets about 2 times higher in 32 nm CMOS. This trend is happening at each technology node and will continue in the future. The implication of such phenomenon is the object of this section.

2.2.1 *MOS Transistor as a Switch*

Figure 2.7 shows the simplified equivalent circuit model of the MOS transistor as a switch in ON and OFF state. The layout parasitics are highlighted in Fig. 2.7b. The main performance parameters of a switch are the ON resistance $R_{ON} \propto 1/g_m$ and OFF capacitance $C_{OFF} \propto C_{GS}$. It is therefore possible to define a figure of merit for the switch as

$$FOM_{SW} = R_{ON}\, C_{OFF} \propto \frac{1}{f_t}. \tag{2.13}$$

Equation 2.13 shows that the transistor switch benefits from technology scaling. However, the effect of the layout parasitics and low level metal interconnects become more important as L_{min} scales, as expected from Fig. 2.6. Figure 2.8 shows that the figure of merit tends to saturate in deep-scaled CMOS, limiting the effective improve of the switch performance when used in a real circuit [4].

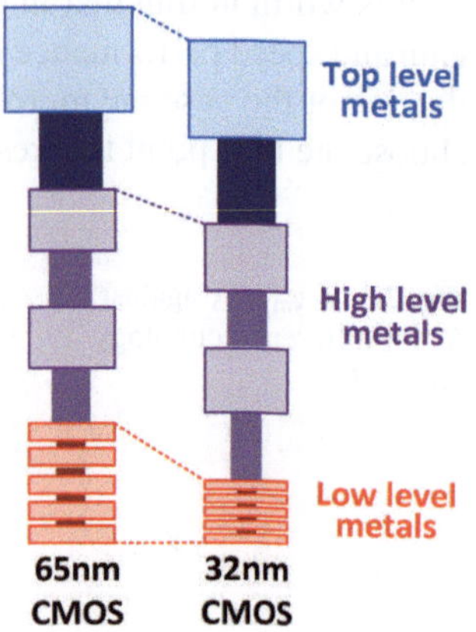

Fig. 2.6 BEOL comparison between 65 and 32 nm CMOS [4]

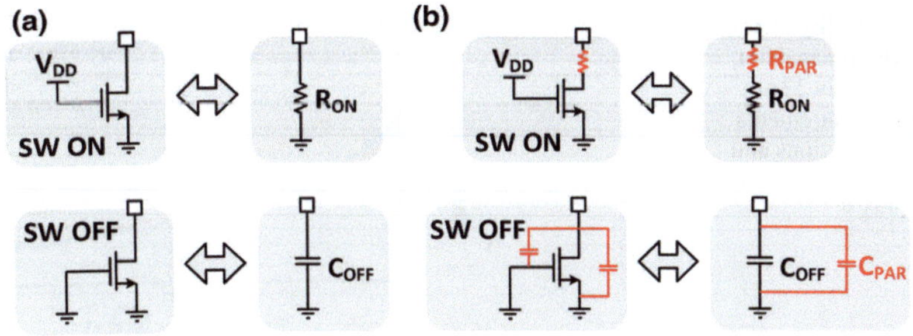

Fig. 2.7 MOS as a switch: equivalent simplified circuits in ON and OFF state, without (**a**) and with (**b**) layout parasitics

Fig. 2.8 MOS switch figure of merit with and without layout parasitics [4]

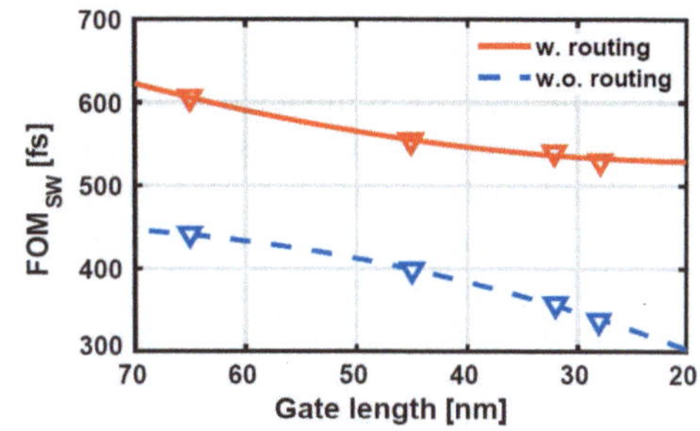

2.2.2 *Capacitors*

Being implemented with the low level metals shown in Fig. 2.6, Metal-Oxide-Metal (MOM) capacitors are the passive components that suffer the most from technology scaling. Moreover, the quality factor of capacitors degrades with frequency

$$Q_C = \frac{1}{2\pi f R_S C}, \tag{2.14}$$

where the capacitor is modeled as an ideal capacitance C in series with a ideal resistor R_S that accounts for the losses. Making the use of such components at mm-Wave not favorable. These effects are clear in Fig. 2.9 [4].

2.2.3 *Inductors*

Inductors are implemented with top metals (Fig. 2.6) to maximize the quality factor and the self-resonant frequency. These components do not benefit from technology scaling, but they do not necessarily degrade either. Moreover, the quality factor of inductors increases with frequency

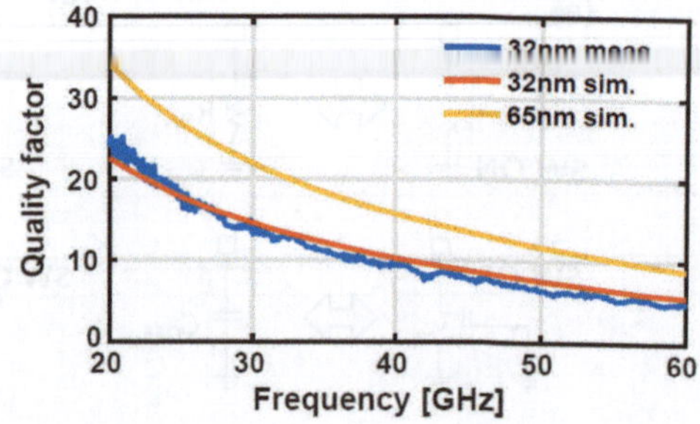

Fig. 2.9 Quality factor of a 250 fF MOM capacitor implemented in 32 nm CMOS. Measurements against simulations and comparison with a 65 nm process [4]

$$Q_L = \frac{2\pi f L}{R_S}, \tag{2.15}$$

where the inductor is modeled as an ideal inductance L in series with a ideal resistor R_S that accounts for the losses. Inductors are largely employed at mm-Wave. The simulated quality factor of a 100 pH inductor implemented in 32 and 65 nm CMOS is shown against frequency in Fig. 2.10 [4].

There are two main technology constrains that limits the practical values of on-chip inductors. (1) Above the self-resonant frequency, the parasitic capacitance to the lossy substrate dominates and the inductor behaves as a capacitors. This effect poses an upper limit to L_{MAX}. It is worth noting that every technology node imposes increasing minimum density rules. Therefore, to pass the design rule check (DRC) on-chip inductors need to be filled with an increasing amount dummies, lowering further the self-resonant frequency [5]. (2) On the other end, the lower bound to the minimum value of L_{min} is set by technology parameters. When L decreases too much, the losses are dominated by VIAs and interconnecting metal, resulting in a dramatic drop of the quality factor [6].

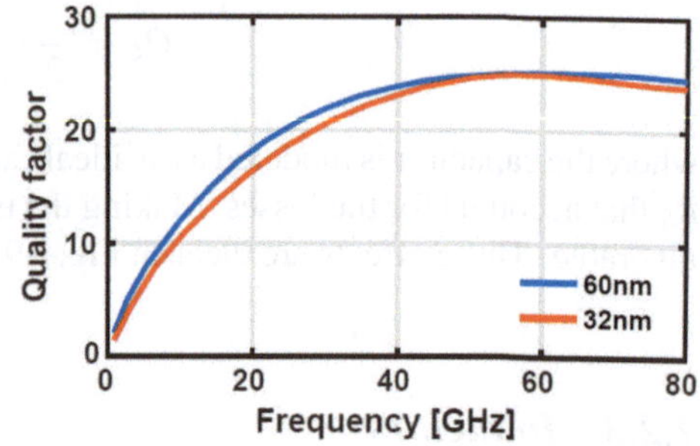

Fig. 2.10 Simulated quality factor of a 100 pH inductor implemented with top metals in 32 versus 65 nm CMOS comparison [4]

2.2.4 Transformers

Two magnetically coupled inductors realize a transformer [7]. The resulting 2-port network schematic is shown in Fig. 2.11 together with three equivalent models. Depending on the specific circuit where the transformer is employed, one of these models leads to an easier analysis.

When the losses are modeled as ideal resistors in series with ideal inductors, the Z-parameter matrix is defined as

$$\begin{bmatrix} V_1 \\ V_2 \end{bmatrix} = \begin{bmatrix} Z_{11} & Z_{12} \\ Z_{21} & Z_{22} \end{bmatrix} \begin{bmatrix} I_1 \\ I_2 \end{bmatrix} = \begin{bmatrix} R_{S,p} + j\omega L_p & j\omega M \\ j\omega M & R_{S,s} + j\omega L_s \end{bmatrix} \begin{bmatrix} I_1 \\ I_2 \end{bmatrix}, \tag{2.16}$$

where $R_{S,p}$, L_p, $R_{S,s}$, L_s are the series resistor and self-inductance of the primary and secondary windings respectively, $M = k\sqrt{L_p L_s}$ and k is the magnetic coupling coefficient.

When the losses are modeled as ideal resistors in parallel with ideal inductors, the Y-parameter matrix is defined as

$$\begin{bmatrix} I_1 \\ I_2 \end{bmatrix} = \begin{bmatrix} Y_{11} & Y_{12} \\ Y_{21} & Y_{22} \end{bmatrix} \begin{bmatrix} V_1 \\ V_2 \end{bmatrix} = \begin{bmatrix} \frac{1}{R_{P,p}} + \frac{1}{j\omega L_p(1-k^2)} & \frac{k}{j\omega\sqrt{(L_p L_s)}(1-k^2)} \\ \frac{k}{j\omega\sqrt{(L_p L_s)}(1-k^2)} & \frac{1}{R_{P,s}} + \frac{1}{j\omega L_s(1-k^2)} \end{bmatrix} \begin{bmatrix} V_1 \\ V_2 \end{bmatrix}, \tag{2.17}$$

where $R_{P,p}$, L_p, $R_{P,s}$, L_s are the parallel resistor and self-inductance of the primary and secondary windings respectively.

On-chip transformers suffer from similar practical limitations as on-chip inductors. The major differences are the following. (1) Due to the parasitic inter-winding capacitance transformers show a lower self-resonant frequency. (2) The magnetic

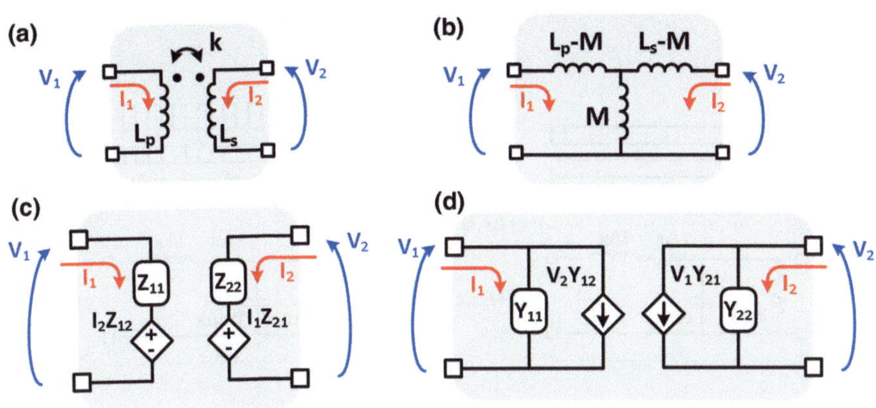

Fig. 2.11 Transformer **a** schematic symbol, **b** equivalent T-section model, **c** Z-parameter and **d** Y-parameter 2-port models

field is more constrained between the two coils, limiting the deleterious effect of dummies when compared to inductors [5, 8].

2.2.5 *Transmission Lines*

Long interconnects carrying mm-Wave signals can be modeled as transmission lines (T-lines). The layout and model of a section of a T-line are shown in Fig. 2.12a. The wave solution is [9]

$$V(x) = V_0^+ e^{-\gamma x} + V_0^- e^{\gamma x}, I(x) = I_0^+ e^{-\gamma x} + I_0^- e^{\gamma x}, \tag{2.18}$$

where the propagation constant γ is

$$\gamma = \alpha + j\beta = \sqrt{(r + j\omega l)(g + j\omega c)}, \tag{2.19}$$

r is the series resistance per unit length [Ω/m], l is the series inductance per unit length [H/m], g is the parallel conductance per unit length [S/m] and c is the parallel capacitance per unit length [F/m] shown in the section model of Fig. 2.12.

It is possible to define the characteristic impedance of a T-line as

$$Z_0 = \sqrt{\frac{r + j\omega l}{g + j\omega c}}. \tag{2.20}$$

The guided wavelength of the signal propagating in the T-line is found as

$$\lambda_g = \frac{2\pi}{Im\{\gamma\}}. \tag{2.21}$$

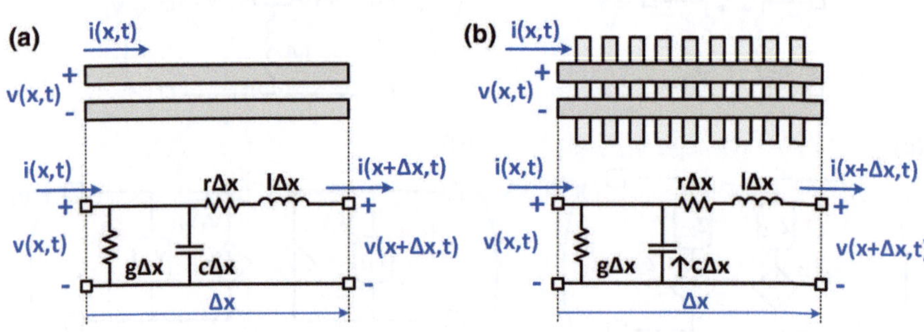

Fig. 2.12 **a** T-line and **b** slow-wave T-line layout and section model

An effective way to reduce the guided wavelength λ_g and reducing the required T-line length (in differential mode) is to place a floating metal shield as shown in Fig. 2.12b [10]. This effect can be intuitively understood as an increased capacitance per unit length. The reduced wavelength results in a slower propagation velocity. This T-line is therefore referred as "slow-wave". Moreover, the shield limits to some extend the losses through the silicon substrate.

The input impedance Z_{IN} of a low loss T-line terminated on a load impedance Z_L can be calculated as

$$Z_{IN} = Z_0 \frac{Z_L + jZ_0 tan(\beta \Delta x)}{Z_0 + jZ_L tan(\beta \Delta x)}, \tag{2.22}$$

where $\beta = \omega\sqrt{lc}$.

Particularly interesting are the following three limit cases. (1) Short, $Z_L = 0$

$$Z_{IN} = jZ_0 tan(\beta \Delta x). \tag{2.23}$$

(2) Open, $Z_L \longrightarrow +\infty$

$$Z_{IN} = \frac{Z_0}{jtan(\beta \Delta x)}. \tag{2.24}$$

(3) Match, $Z_L = Z_0$

$$Z_{IN} = Z_0. \tag{2.25}$$

The input impedance in Eq. 2.23 behaves as an inductor and in Eq. 2.24 as a capacitor.[1] The major difference with a lumped element realization is that the values of this inductor/capacitor varies periodically with frequency, through the dependency on β. Equation 2.25 shows that when a TL is matched to the load impedance, $Z_{IN} = Z_0$ for any frequency. In the case of a low loss TL, Z_0 is a real number therefore the load impedance should be perfectly resistive to realize ideal matching at every frequency.

Other limit cases of interest for some applications are the following. Half wave length T-line, $tan(\beta \Delta x) = 0$ (i.e. $\beta \Delta x = n\pi$ or $\Delta x = n\lambda_g/2$)

$$Z_{IN} = Z_L. \tag{2.26}$$

Quarter wave length TL, $tan(\beta \Delta x) \longrightarrow \pm\infty$ (i.e. $\beta \Delta x = n\pi/2$ or $\Delta x = n\lambda_g/4$)

$$Z_{IN} = \frac{Z_0^2}{Z_L}. \tag{2.27}$$

This last expression shows that a quarter wavelength T-line behaves as an impedance inverter. If Z_L increases, Z_{IN} decreases and vice versa. Such an element is key to realize the load modulation effect in an ideal Doherty power amplifier [11, 12].

---

[1] As long as $tan(\beta \Delta x) > 0$. This is always the case when $0 < \Delta x < \lambda_g/4$. This condition is desirable since the silicon area of a TL is substantial.

Inductors are often used in mm-Wave design to resonate the parasitic capacitance of the active devices or to realize a tank in an LC oscillators. When an inductor is realized with a T-line, it typically shows lower losses and better return path modeling, making the design scalable and drastically reducing the EM simulations needed to synthesis the required inductance [13]. However, lumped inductors feature a lower silicon area consumption and enable on-chip transformers, providing galvanic isolation and easing the DC bias feed to the circuitry. Therefore, in this work lumped element components are preferred whenever possible.

2.3 Conclusion

This chapter has focused on the major implications of CMOS technology scaling on the design of active and passive devices for mm-Wave applications. Every circuit building block discussed in the following chapters makes use of the insights and design equations derived here.

The transistor operation as a transconductance amplifier has been discussed in detail. The inversion coefficient (IC) has been introduced, resulting in simple design guidelines to achieve maximum gain and speed for given power consumption. The IC sheds unique insight on the evolution of mm-Wave analog design from the past to the present and allows a qualitative prediction for the evolution in the near future.

The effect of scaling on passive devices has been investigated. Interestingly, it has been shown that CMOS scaling does not provide any obvious benefit. When the intrinsic MOS device with a smaller feature side is used as a switch, it shows an higher figure of merit. However, the effect of the low level metal connections tends to cancel out the benefit. The quality factor of capacitors gets worse and inductors do not improve.

References

1. W.M.C. Sansen, *Analog Design Essentials*, vol. 859 (Springer Science and Business Media, Berlin, 2007)
2. C.C. Enz, E.A. Vittoz, *Charge-Based MOS Transistor Modeling: The EKV Model for Low-Power and RF IC Design* (Wiley, New York, 2006)
3. W. Sansen, 1.3 Analog CMOS from 5 micrometer to 5 nanometer, in *2015 IEEE International Solid-State Circuits Conference - (ISSCC) Digest of Technical Papers, San Francisco, CA* (2015), pp. 1–6
4. E. Mammei, E. Monaco, A. Mazzanti, F. Svelto, A 33.6-to-46.2 GHz 32 nm CMOS VCO with 177.5 dBc, Hz minimum noise FOM using inductor splitting for tuning extension, in *2013 IEEE International Solid-State Circuits Conference Digest of Technical Papers, San Francisco, CA* (2013), pp. 350–351
5. D. Zhao, P. Reynaert, A 60 GHz dual-mode class AB power amplifier in 40 nm CMOS. IEEE J. Solid-State Circuits **48**(10), 2323–2337 (2013)

6. S.A.R. Ahmadi-Mehr, M. Tohidian, R.B. Staszewski, Analysis and design of a multi-core oscillator for ultra-low phase noise. IEEE Trans. Circuits Syst. I Regul. Pap. **63**(4), 529–539 (2016)
7. J.R. Long, Monolithic transformers for silicon RF IC design. IEEE J. Solid-State Circuits **35**(9), 1368–1382 (2000)
8. F.-W. Kuo et al., A 12 mW all-digital PLL based on class-F DCO for 4G phones in 28 nm CMOS, *2014 Symposium on VLSI Circuits Digest of Technical Papers, Honolulu, HI* (2014), pp. 1–2
9. D.M. Pozar, *Microwave Engineering* (Wiley, New York, 2009)
10. T.S.D. Cheung, J.R. Long, Shielded passive devices for silicon-based monolithic microwave and millimeter-wave integrated circuits. IEEE J. Solid-State Circuits **41**(5), 1183–1200 (2006)
11. W.H. Doherty, A new high efficiency power amplifier for modulated waves. Proc. Inst. Radio Eng. **24**(9), 1163–1182 (1936)
12. S.C. Cripps, Inverted logic [Microwave Bytes]. IEEE Microw. Mag. **9**(5), 30–38 (2008)
13. A. Mazzanti, M. Sosio, M. Repossi, F. Svelto, A 24 GHz subharmonic direct conversion receiver in 65 nm CMOS. IEEE Trans. Circuits Syst. I Regul. Pap. **58**(1), 88–97 (2011)

Chapter 3
Gain-Bandwidth Enhancement Techniques for mm-Wave Fully-Integrated Amplifiers

This chapter recalls filter basics and introduces design techniques to achieve gain-bandwidth enhancement and further approach the Bode-Fano limit. A strong focus is put on filter topologies that lead to relatively easy implementation with on-chip components and have shown state-of-the-art performance at mm-Wave.

Section 3.1 discusses the basic RLC band-pass filter. Filter quality factor and noise are briefly recalled, setting the foundation of resonant circuits for mm-Wave application for both amplifiers and oscillators. Section 3.2 introduces 4th order filters designed to achieve gain-bandwidth enhancement when compared to the classical RLC tank. Several topologies are discussed and compared. Simple design equations are derived. Transformer based resonators are the focus of Sect. 3.3. The effect of the parasitic interwinding capacitance is discussed, providing intuition on the circuit operation and simple design guidelines. Next, the discussion is extended to achieve impedance transformation and realize power dividers and combiners.

3.1 RLC Tank

3.1.1 RC Low-Pass Filter

The basic RC low-pass filter schematic is shown in Fig. 3.1. The admittance of the circuit is

$$Y = sC + \frac{1}{R} = \frac{sRC + 1}{R}. \tag{3.1}$$

The impedance is simply $Z = 1/Y$, and it shows a low-pass behavior with a single pole at $\omega_p = 1/(RC)$.

M. Vigilante and P. Reynaert, *5G and E-Band Communication Circuits in Deep-Scaled CMOS*, Analog Circuits and Signal Processing,
https://doi.org/10.1007/978-3-319-72646-5_3

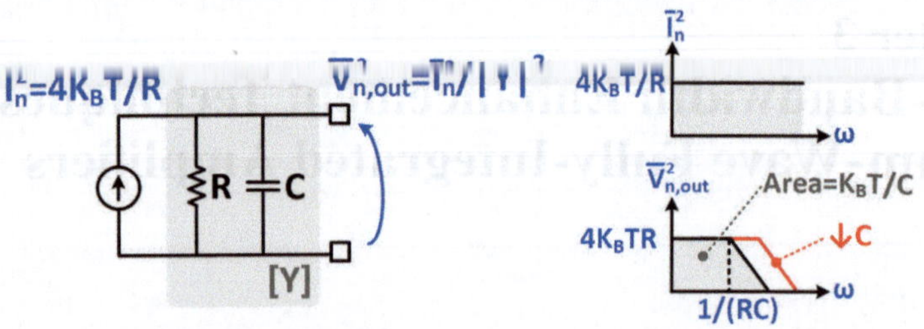

Fig. 3.1 *RC* low-pass filter schematic and noise

The resistor in the network is responsible for thermal noise [1]

$$\overline{I}_n^2 = 4K_B\,T/R. \tag{3.2}$$

This noise current produces an output noise $\overline{V}_{n,out}^2$ voltage that is shaped by the filter transfer function, as shown in Fig. 3.1. The higher the filter capacitance the lower the total integrated noise $K_B\,T/C$ [1].

The quality factor of the filter is

$$Q = \frac{Im\{Y\}}{Re\{Y\}} = \frac{Im\{Z\}}{Re\{Z\}} = \omega RC. \tag{3.3}$$

3.1.2 RLC Band-Pass Filter

By adding an inductor L, the low-pass RC filter is upconverted to $\omega_o = 1/(\sqrt{LC})$. Figure 3.2 shows the resulting RLC band-pass filter and its output noise. Intuitively, at f_o the inductor and the capacitor resonate and the tank reduces to R_T. The output noise at ω_o therefore is simply [2]

$$\overline{V}_n^2(\omega_o) = \frac{4K_B\,T}{R_T} R_T^2 = 4K_B\,T\,R_T. \tag{3.4}$$

The total integrated noise still reduces with $1/C$. The quality factor of the filter is

$$Q = \frac{Im\{Z\}}{Re\{Z\}} = \omega_o\,R_T C. \tag{3.5}$$

This simple example is of utmost importance. (1) An oscillator can be effectively modeled in steady-state as a current source in parallel with an RLC tank [3], even when a 4th tank is used [4]. It is therefore reasonable to expect that a tank with a higher Q-factor results in a benefit it terms of noise. And (2), an amplifier can be effectively modeled as a voltage driven current source with a parallel RC output impedance

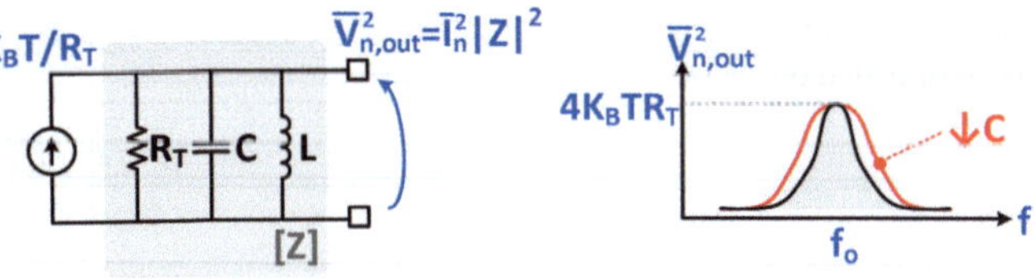

Fig. 3.2 *RLC* band-pass filter schematic and noise

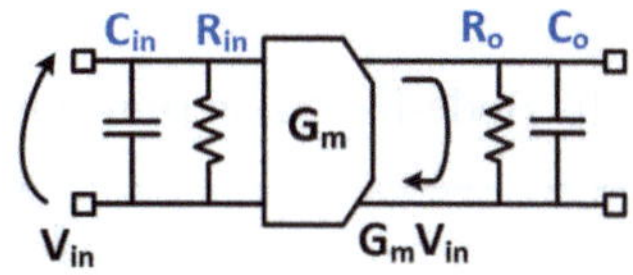

Fig. 3.3 mm-Wave CMOS amplifier simplified circuit model

(Fig. 3.3). By adding an inductor, signal amplification at mm-Wave becomes possible. In this second case the load RC product limits the achievable $-3\,\text{dB}$ bandwidth ($BW_{-3\,\text{dB}}$). The only viable way to enlarge the bandwidth in this case is adding and equivalent parallel resistor, lowering the tank Q-factor (in Eq. 3.5) at the cost of extra losses (i.e. lower efficiency and higher noise).

3.2 Coupled Resonators

3.2.1 Bode-Fano Limit

A typical mm-Wave CMOS amplifier can be modeled as an ideal transconductance G_m with a parallel RC input and output impedance, $R_{in}//C_{in}$ and $R_o//C_o$ respectively (as shown in Fig. 3.3). To achieve the required gain, noise, input match and/or output power specifications, filters are needed to resonate the parallel capacitance and realize impedance transformation over the required bandwidth. In this sense, low-noise amplifiers, power amplifiers, on-chip gain stages and buffers share to some extend similar design challenges and solutions.

As shown in Sect. 3.1 a simple way to resonate the capacitor is to add an inductor L in parallel. This leads to a $BW_{-3\,\text{dB}}$ limited by the RC product of the tank. A few questions rise. (1) Is it possible to perfectly resonate C over a large bandwidth? (2) Does it exist a theoretical limit to this problem? (3) Does it exist a theoretical optimum solution?

The Bode-Fano limit answers these questions [5, 6]. In Fig. 3.4 an ideal passive lossless filter is terminated on a parallel RC load. The input impedance of the filter is Z_{in}. The reflection coefficient measures how close the input impedance is to the resistive termination R over the frequency range (i.e. how well the load capacitance is canceled over frequency)

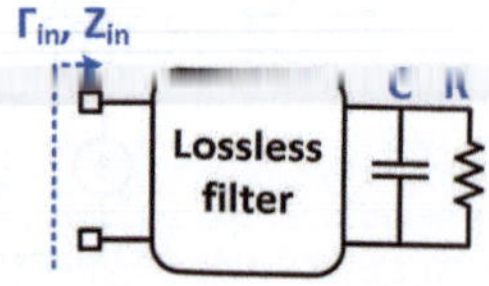

Fig. 3.4 Bode-Fano limit for a lossless filter terminated on a parallel RC load

$$\Gamma_{in}(f) = \frac{Z_{in} - R}{Z_{in} + R}. \tag{3.6}$$

The Bode-Fano criterion states that

$$\int_0^{+\infty} ln\left(\frac{1}{|\Gamma_{in}(\omega)|}\right) d\omega \leq \frac{\pi}{RC}. \tag{3.7}$$

It is worth noting that the magnitude of the reflection coefficient and the in-band ripple are closely related [7]

$$|\Gamma_{in}| = \sqrt{1 - \frac{1}{Ripple}}. \tag{3.8}$$

The implications of this simple result are the following. (1) For a given RC load, a broader pass-band bandwidth can be achieved only at expenses of larger ripple. (2) The capacitance C can be resonated out perfectly only at a finite number of frequencies. (3) High-Q circuits are more difficult to match than low-Q ones. (4) For a given finger length, the RC product of a transistor does not vary with its width for a given technology. Making the results of this simple analysis extremely general. For example, a low noise amplifier with $W_{LNA} = 20\,\mu\text{m}$ shows exactly the same Q-factor of a power amplifier $W_{PA} = 5 \cdot 40\,\mu\text{m} = 200\,\mu\text{m}$. This is the case since a wider transistor can be realize by using more fingers in parallel and/or by using multiple transistors in parallel. Leading to and increase of the equivalent parallel input/output capacitance proportional to W and a decrease of the equivalent parallel R_{in}, R_o proportional to $1/W$.

The ideal pass-band filter can only be approximated in a real implementation. A close practical approximation is the Chebyshev filter [8]. However, high order filters demands a large number of passive components. Given the technology constrains discussed in Chap. 2, the effectiveness of such techniques has been limited so far to the low GHz range [7, 9, 10].

At mm-Wave frequencies 4th order coupled resonators offer gain-bandwidth enhancement when compared to the simple RLC tank, without jeopardizing the network efficiency. Therefore, the rest of this work is focus on these kinds of filters.

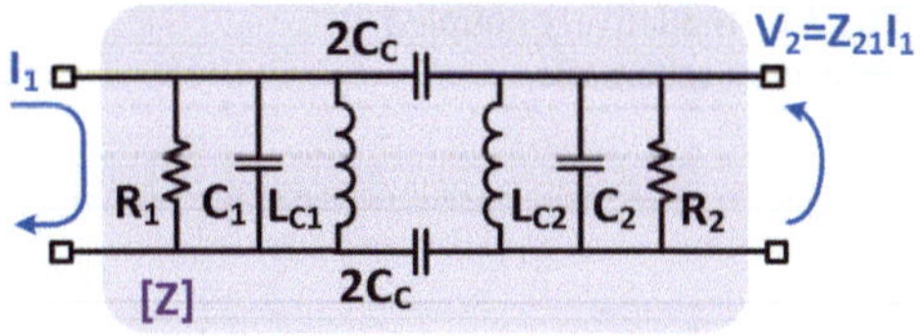

Fig. 3.5 Capacitively coupled resonators schematic

3.2.2 *Capacitively Coupled Resonators*

Two RLC resonators can be coupled by means of a capacitor C_C as shown in Fig. 3.5 [11]. When $R_1 = R_2 = R$, $C_1 = C_2 = C$ and $L_{C1} = L_{C2} = L_C$, the admittance parameters of this two-port network are

$$Y_{11} = Y_{22} = \frac{1}{R} + \frac{1}{sL_C} + s(C + C_C) = \frac{1}{R}\left[1 + Q\left(\frac{s}{\omega_o} + \frac{\omega_o}{s}\right)\right], \tag{3.9}$$

$$Y_{21} = Y_{12} = -sC_C = -\frac{s\,k_C\,Q}{\omega_o\,R}, \tag{3.10}$$

where

$$\omega_o = \frac{1}{\sqrt{L_C(C + C_C)}}, \tag{3.11}$$

$$Q = \frac{R}{\omega_o L_C} = \omega_o R(C + C_C), \tag{3.12}$$

$$k_C = \frac{C_C}{C + C_C}. \tag{3.13}$$

The transimpedance of the two-port network can be found as [8, 11]

$$Z_{21} = \frac{-Y_{21}}{Y_{11}Y_{22} - Y_{12}Y_{21}} = \\ = \frac{s^3 k_C Q \omega_o R}{[Q(1 + k_C)s^2 + s\omega_o + Q\omega_o^2][Q(1 - k_C)s^2 + s\omega_o + Q\omega_o^2]}. \tag{3.14}$$

Assuming high quality factor, the two complex poles of Z_{21} can be calculated as

$$\omega_L = \frac{1}{\sqrt{L_C(C + 2C_C)}}, \tag{3.15}$$

$$\omega_H = \frac{1}{\sqrt{L_C\,C}}. \tag{3.16}$$

A larger C_C allows a larger the band-pass bandwidth at the expenses of an increased quality factor of the network (see Eq. 3.12) and in-band ripple.

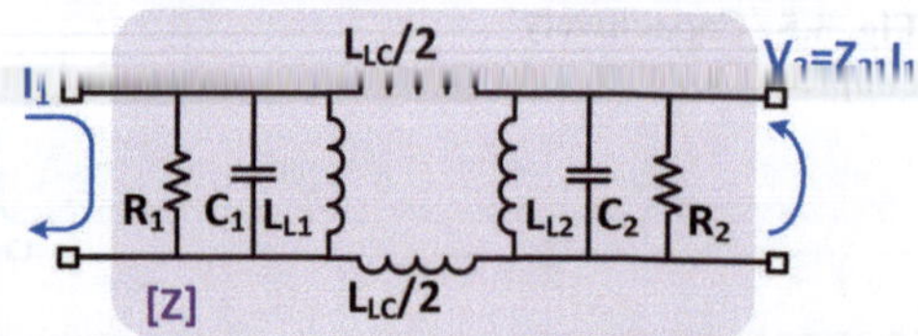

Fig. 3.6 Inductively coupled resonators schematic

3.2.3 Inductively Coupled Resonators

Two RLC tanks can be coupled through an inductor L_c. The schematic of the resulting filter is shown in Fig. 3.6 [12, 13]. When $R_1 = R_2 = R$, $C_1 = C_2 = C$ and $L_{L1} = L_{L2} = L_L$, the admittance parameters of this two-port network can be written as

$$Y_{11} = Y_{22} = \frac{1}{R} + \frac{L_{LC} + L_L}{s\,L_L\,L_{LC}} + sC = \frac{1}{R}\left[1 + Q\left(\frac{s}{\omega_o} + \frac{\omega_o}{s}\right)\right], \tag{3.17}$$

$$Y_{21} = Y_{12} = -\frac{1}{sL_{LC}} = -\frac{\omega_o\,k_L\,Q}{s\,R}, \tag{3.18}$$

where

$$\omega_o = \frac{1}{\sqrt{\frac{L_L\,L_{LC}}{L_L + L_{LC}}C}}, \tag{3.19}$$

$$Q = \frac{R(L_L + L_{LC})}{\omega_o L_L\,L_{LC}} = \omega_o R\,C, \tag{3.20}$$

$$k_L = \frac{L_L}{L_{LC} + L_L}. \tag{3.21}$$

The transimpedance of the two-port network can be found as

$$Z_{21} = \frac{\omega_o^3 k_L\,Q\,R\,s}{[Q\,s^2 + s\omega_o + Q(1 + k_L)\omega_o^2][Q\,s^2 + s\omega_o + Q(1 - k_L)\omega_o^2]}. \tag{3.22}$$

Assuming high quality factor, the two complex poles of Z_{21} can be calculated as

$$\omega_L = \frac{1}{\sqrt{L_L\,C}}, \tag{3.23}$$

$$\omega_H = \frac{1}{\sqrt{\frac{L_L\,L_{LC}}{2L_L + L_{LC}}C}}. \tag{3.24}$$

By selecting a lower value of L_{LC} a larger band-pass bandwidth can be achieved at the expenses of an increased in-band ripple.

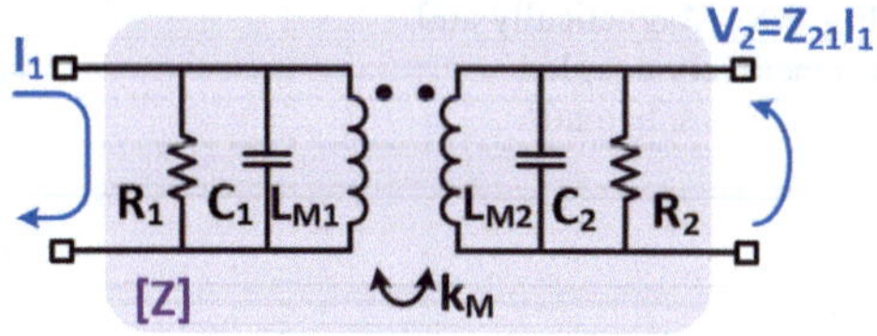

Fig. 3.7 Magnetically coupled resonators schematic

3.2.4 *Magnetically Coupled Resonators*

Two RLC tanks can be magnetically coupled by means of a transformer. The schematic of the resulting filter is shown in Fig. 3.7 [14]. $R_1 = R_2 = R$, $C_1 = C_2 = C$ and $L_{M1} = L_{M2} = L_M$ are assumed in the following. When the Y-parameter model of the transformer in Eq. 2.17 is adopted, it is straightforward to derive the admittance parameters of this two-port network

$$Y_{11} = Y_{22} = \frac{1}{R} + \frac{1}{sL_M(1-k_M^2)} + sC = \frac{1}{R}\left[1 + Q\left(\frac{s}{\omega_o} + \frac{\omega_o}{s}\right)\right], \tag{3.25}$$

$$Y_{21} = Y_{12} = \frac{k_M}{sL_M(1-k_M^2)} = \frac{k_M\,\omega_o\,Q}{s\,R}, \tag{3.26}$$

where

$$\omega_o = \frac{1}{\sqrt{L_M(1-k_M^2)C}}, \tag{3.27}$$

$$Q = \frac{R}{\omega_o L_M(1-k_M^2)} = \omega_o R\,C. \tag{3.28}$$

The transimpedance of the two-port network can be found as

$$Z_{21} = \frac{-\omega_o^3 k_M\,Q\,R\,s}{[Q\,s^2 + s\omega_o + Q(1+k_M)\omega_o^2][Q\,s^2 + s\omega_o + Q(1-k_M)\omega_o^2]}. \tag{3.29}$$

Assuming high quality factor, the two complex poles of Z_{21} can be calculated as

$$\omega_L = \frac{1}{\sqrt{L_M(1+|k_M|)C}}, \tag{3.30}$$

$$\omega_H = \frac{1}{\sqrt{L_M(1-|k_M|)C}}. \tag{3.31}$$

A larger magnetic coupling coefficient k allows a larger band-pass bandwidth at the expenses of an increased Q-factor of the filter (see Eq. 3.28) and in-band ripple.

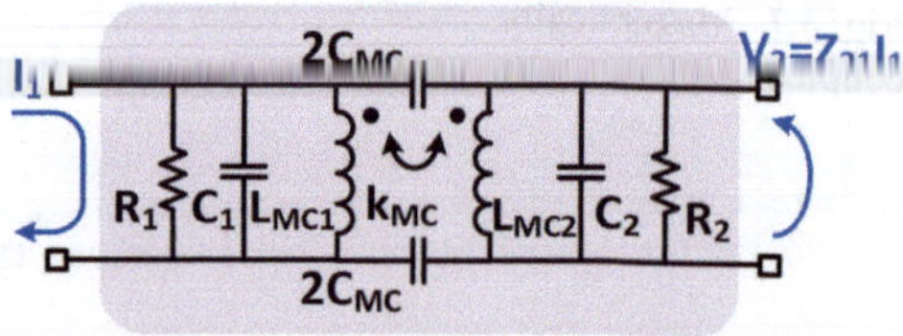

Fig. 3.8 Magnetically and Capacitively coupled resonators schematic

3.2.5 *Magnetically and Capacitively Coupled Resonators*

It is possible to couple two RLC tanks both magnetically and capacitively, as shown in Fig. 3.8 [15]. Once again, the analysis can be greatly simplified when the Y-parameter model is adopted and $R_1 = R_2 = R$, $C_1 = C_2 = C$ and $L_{MC1} = L_{MC2} = L_{MC}$ are assumed. By inspection, the admittance parameters of this filter are derived as follow

$$\begin{aligned} Y_{11} = Y_{22} &= \frac{1}{R} + \frac{1}{sL_{MC}(1-k_{MC}^2)} + s(C + C_{MC}) = \\ &= \frac{1}{R}\left[1 + Q\left(\frac{s}{\omega_o} + \frac{\omega_o}{s}\right)\right], \end{aligned} \tag{3.32}$$

$$Y_{21} = Y_{12} = -sC_{MC} - \frac{k_{MC}}{sL_{MC}(1-k_{MC}^2)} = -\frac{Q}{R}\left(\frac{sk_C}{\omega_o} + \frac{k_{MC}\omega_o}{s}\right), \tag{3.33}$$

where

$$\omega_o = \frac{1}{\sqrt{L_{MC}(1-k_{MC}^2)(C + C_{MC})}}, \tag{3.34}$$

$$Q = \frac{R}{\omega_o L_{MC}(1-k_{MC}^2)} = \omega_o R\,(C + C_{MC}), \tag{3.35}$$

$$k_C = \frac{C_{MC}}{C + C_{MC}}. \tag{3.36}$$

The transimpedance of the two-port network can be found as

$$Z_{21} = \frac{\omega_o\, Q\, R\, s(k_C\, s^2 + k_{MC}\, \omega_o^2)}{Den}, \tag{3.37}$$

$$\begin{aligned} Den = {} & [Q(k_C - 1)s^2 + s\omega_o + Q(1 - k_{MC})\omega_o^2]\cdot \\ & \cdot [Q(k_C + 1)s^2 + s\omega_o + Q(1 + k_{MC})\omega_o^2]. \end{aligned} \tag{3.38}$$

Assuming high quality factor and $k_{MC} < 0$, the two complex poles of Z_{21} can be calculated as

$$\omega_L = \frac{1}{\sqrt{L_{MC}(C + 2C_{MC})(1 - k_{MC})}}, \tag{3.39}$$

$$\omega_H = \frac{1}{\sqrt{L_{MC}(1 + k_{MC})C}}. \tag{3.40}$$

A larger band-pass bandwidth can be achieved at the expenses of an increased in-band ripple by increasing k_{MC} and C_{MC}.

3.2.6 *Coupled Resonators Comparison*

The best way to compare the aforementioned 4th order filters is to consider a simple design example. Typical values are adopted for the input and output impedances of a G_m stage implemented in 28 nm bulk CMOS shown in Fig. 3.3. $R_1 = R_o = 400\Omega$, $R_2 = R_{in} = 1\,\mathrm{k}\Omega$ and $C = C_1 = C_o = C_2 = C_{in} = 14fF$. The filters are designed to achieve roughly the same >30% fractional bandwidth around the center frequency $f_o = 80$ GHz, resulting in >24 GHz $BW_{-3\,\mathrm{dB}}$. $\omega_L = 2\pi\,68$ GHz and $\omega_H = 2\pi\,92$ GHz are imposed. The filter based on magnetically and capacitively coupled resonators (Fig. 3.8) can be designed to equalize the magnitude of the filter transimpedance at the two maxima by further imposing the conditions $k_{MC} < 0$ and $C_C = -C\,k_{MC}/(1 + k_{MC})$ as proposed in [15].

The result of this investigation is shown in Fig. 3.9 together with the transimpedance Z_{21} of a classical tuned transformer with $k = 0.8$ for comparison. Clearly, 4th order filters show a gain-bandwidth enhancement when compared to a simple RLC tank or a tuned transformer. When the latter is considered, the only way to achieve a larger bandwidth is to lower the quality factor of the load by adding an equivalent parallel resistor, compromising the insertion loss of the filter.

Inductively coupled and magnetically coupled resonators stand out for the lowest ripple for a given bandwidth (Fig. 3.9). Capacitively coupled resonators are the furthest from the Bode-Fano limit. Perhaps not surprisingly, filters based on both capacitive and magnetic coupling achieve performance in between the two. This can

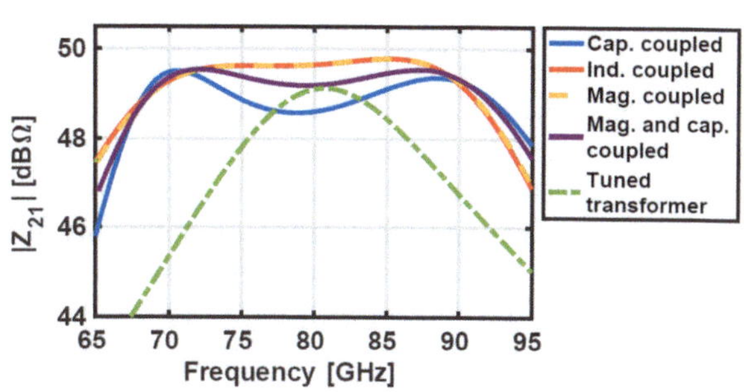

Fig. 3.9 Comparison of the transimpedance frequency response of different 4th order filters. ©2017 IEEE. Reprinted, with permission, from [16]

be intuitively understood as follow. The quality factor of a filter is proportional to the RC constant of the load (Eqs. 3.3, 3.5, 3.12, 3.20, 3.28 and 3.35) and the higher the Q the larger the ripple, according to the Bode-Fano limit (see Eq. 3.7). It is therefore not desirable to add capacitance to the network. Moreover, on-chip capacitors at mm-Wave suffers from a reduced quality factor, that degrades every technology node (Sect. 2.2).

It is worth noting that in a practical on-chip implementation a transformer allows a substantial area reduction, easier DC feed and AC coupling when compared to the inductively coupled resonators. The two filters show exactly the same $|Z_{21}|$ in Fig. 3.9 as expected from Eqs. 3.22 and 3.29. This can be intuitively understood by noting that a transformer can be equivalently modeled as a Π-network composed of three inductors single-ended or four inductors differentially. However, on-chip inductors couple to and from other circuits. Whereas, the magnetic field is better constrained in a transformer, making the coupling easier to control and model. Finally, the impact of dummies in the final layout is also less critical for a transformer, as discussed in Sect. 2.2.

For all these reasons, transformer-based filters are considered hereafter.

3.3 Transformer-Based Resonators

3.3.1 *On the Parasitic Interwinding Capacitance*

Figure 3.10 shows the layout of typical inverting and non-inverting 1:1 transformers and their equivalent lumped element models [17]. Highlighted in gray in Fig. 3.10c are the parasitics to the silicon substrate C_{ox}, C_{Si}, r_{Si}, the parasitic intra-winding capacitance C_{m1}, C_{m2} and the inter-winding capacitance C_C. Even if this model is rather accurate over a very large bandwidth [18], due to its complexity, it is particularly involved to extract the exact values of each component in the schematic from measurements and simulation. Making it really challenging to develop a scalable model. Therefore, designers need to largely rely on electromagnetic (EM) simulators to accurately describe this network [2].

Nevertheless, it is instructive to focus on the simplified differential mode (DM) equivalent model in Fig. 3.10d. The parasitics to the substrate and the intra-winding capacitance are modeled as an equivalent parallel RC network, making it possible to absorb them in the filter terminations. To further simplify the analysis and get insight into the effect of the parasitic interwinding capacitance C_C, in the following R_{S1} and R_{S2} are neglected. The schematic of the resulting 2-port filter is shown in Fig. 3.10d.

Intuitively, we expect that if the current flowing through the parasitic interwinding capacitance is $I_{Cc} = 0$, C_C has no effect on the filter frequency response. This happens when the voltage across C_C, $V_{Cc} = 0$. By the same token, when I_{Cc} is maximum, the effect of C_C is also maximize. Interestingly, the voltage across C_C can be written as

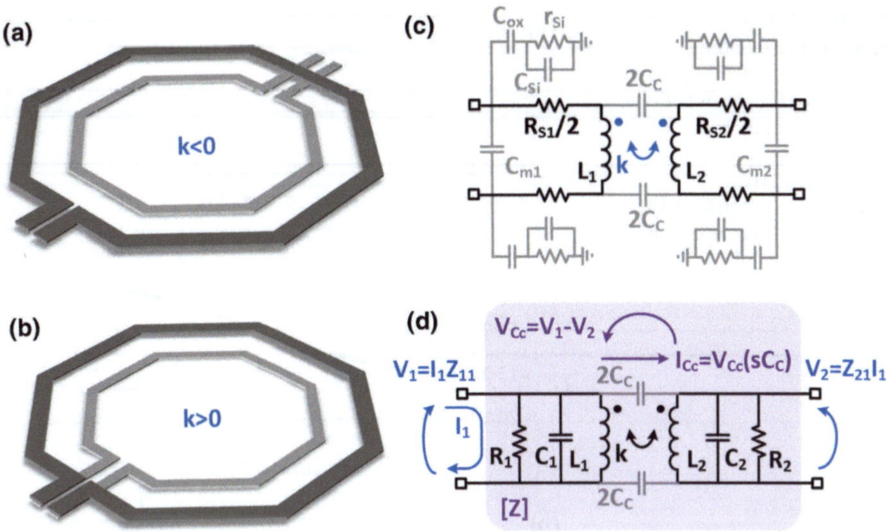

Fig. 3.10 Layout example of an inverting transformer (**a**) and a non-inverting one (**b**), along with its equivalent lumped element model including layout parasitic (**c**) and its simplified schematic in DM (**d**). ©2017 IEEE. Reprinted, with permission, from [16]

$$
\begin{aligned}
V_{Cc} = V_1 - V_2 = I_1(Z_{11} - Z_{21}) &= I_1\left(|Z_{11}|\, e^{j\angle Z_{11}} - |Z_{21}|\, e^{j\angle Z_{21}}\right) \\
&= I_1\, |Z_{21}| e^{j\angle Z_{21}} \left(\frac{|Z_{11}|}{|Z_{21}|}\, e^{j(\angle Z_{11} - \angle Z_{21})} - 1\right).
\end{aligned}
\tag{3.41}
$$

Equation 3.41[1] shows that regardless the magnitude of Z_{11} and Z_{21}, the voltage drop across C_C is maximum when $\angle Z_{11} - \angle Z_{21} = \pm 180°$ and minimum when $\angle Z_{11} - \angle Z_{21} = 0°$.

This insight is key to understand the effect of C_C on the filter response. To further investigate this parasitic effect, let us go back to the previous design example and assume a parasitic inter-winding capacitance $C_C = 1\, fF$. This value is reasonable for a $f_o \approx 80$ GHz center frequency when relatively low-k transformers are used, and it is optimistic when high-k transformers are designed (as it is the case for a classical tuned transformer). When C_C is neglected, the sign of the magnetic coupling coefficient k has no effect on the magnitude of the transimpedance of the filter. The $BW_{-3\,\mathrm{dB}} \approx 31.3$ GHz with $\approx$0.16 dB in-band ripple. However, when the parasitic inter-winding capacitance is considered, inverting and not-inverting transformers behave very differently. This is clearly shown in Fig. 3.11. An inverting transformer ($k < 0$) realizes $\angle Z_{11} - \angle Z_{21}$ that goes from $\approx -150°$ in the proximity of the

[1] It is worth noting that in general the voltage at port 1 and port 2 in a 2-port network can be written as $V_1 = Z_{11}I_1 + Z_{12}I_2$ and $V_2 = Z_{21}I_1 + Z_{22}I_2$, respectively. However, in the case under discussion the 2-port filter is terminated on an open circuit, i.e. $I_2 = 0$. Therefore, in this case $V_1 = Z_{11}I_1$ and $V_1 = Z_{21}I_1$. Equation 3.41 directly follows.

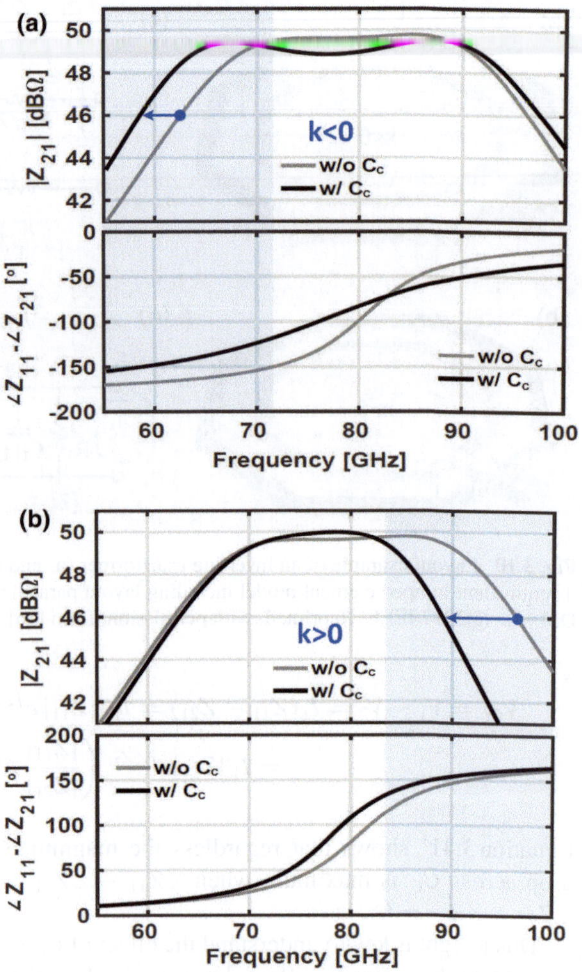

Fig. 3.11 Effect of the parasitic inter winding capacitance on the frequency response of broadband 4th order filters implemented with inverting (**a**) and non-inverting (**b**) transformers. ©2017 IEEE. Reprinted, with permission, from [16]

low frequency pole f_L to $\approx -10°$ in the proximity of the high frequency pole f_H. The resulting effect is that C_C lowers f_L while keeping f_H unchanged, realizing a wider bandwidth ($BW_{-3\,\text{dB}} \approx 37.3$ GHz) with larger ripple (≈0.67 dB) as shown in Fig. 3.11a. The contrary is happening when the same filter uses a non-inverting transformer ($k > 0$), Fig. 3.11b. In this second case f_H is moved towards lower frequencies, while f_L does not change. This results in ≈9 GHz lower bandwidth.

It is worth noting that in a broadband design when a transformer with $k < 0$ is adopted, to counter act the effect of C_C, a lower L_1, L_2 and k can be used. This results in a further reduction of the parasitic inter-winding capacitance. When a non-inverting transformer is used, however, to counter act the effect of C_C a larger value of k is needed. This normally results in a further enhancement of the parasitic inter-winding capacitance. Therefore, it is desirable to use inverting transformers for this kind of networks whenever possible.

Finally, assuming $L_1 = L_2 = L$ (as in the example considered so far), the self-resonant frequency of the transformer can be derived as

$$f_{SRF} = \frac{1}{2\pi\sqrt{2L(1-k)C_C}}. \tag{3.42}$$

It is now possible to compare the self-resonant frequencies of an inverting ($f_{SRF,k<0}$) and non-inverting transformer ($f_{SRF,k>0}$) with same L, C_C and $|k|$

$$f_{SRF,k<0} = \sqrt{\frac{(1-|k|)}{(1+|k|)}}\, f_{SRF,k>0}. \tag{3.43}$$

Interestingly, $f_{SRF,k<0}$ is always lower than $f_{SRF,k>0}$. Therefore, when the f_{SRF} is adopted as a figure of merit to benchmark the upper frequency limit of a transformer, it may lead to the wrong conclusion that $k > 0$ is always preferable when circuits with high frequency of operation are designed.

3.3.2 *Effect of Unbalanced Capacitive Terminations*

The expressions of ω_L and ω_H have been derived assuming $C_1 = C_2 = C$. In general this is not the case and it is interesting to see how the different filters previously introduced respond to this effect. Following the same approach outlined in [11–13, 15] and carrying out the algebra, the frequency of the complex poles when $C_2 = n\,C_1$ can be derived as

$$\begin{aligned}\omega_L &= \frac{1}{\sqrt{L_{C1}(C_1 + C_C(1+1/n))}} = \frac{1}{\sqrt{L_{L1}\,C_1}} = \frac{1}{\sqrt{L_{L2}\,C_2}}\\ &= \frac{1}{\sqrt{L_{M1}(1+|k_M|)C_1}} = \frac{1}{\sqrt{L_{M2}(1+|k_M|)C_2}},\end{aligned} \tag{3.44}$$

$$\begin{aligned}\omega_H &= \frac{1}{\sqrt{L_{C1}\,C_1}} = \frac{1}{\sqrt{L_{C2}\,C_2}} = \frac{1}{\sqrt{\frac{L_{L1}\,L_{LC}}{L_{L1}(1+1/n)+L_{LC}}C_1}}\\ &= \frac{1}{\sqrt{L_{M1}(1-|k_M|)C_1}} = \frac{1}{\sqrt{L_{M2}(1-|k_M|)C_2}}.\end{aligned} \tag{3.45}$$

Unfortunately, the close form expressions for the filter in Fig. 3.8 do not lead to simple design guidelines or insight, therefore have been omitted.

Figure 3.12 shows the effect of $n = 2$ on the frequency response of the filters. This is a typical value for the inter-stage matching network of a power amplifier, where the driver is downsized by a factor of 2, and a somewhat extreme case for an LNA were normally the size of the amplifiers in the chain is not increased to save power [11, 13]. $\omega_L = 2\pi\,68\,\text{GHz}$ and $\omega_H = 2\pi\,92\,\text{GHz}$ are imposed in Eqs. 3.44

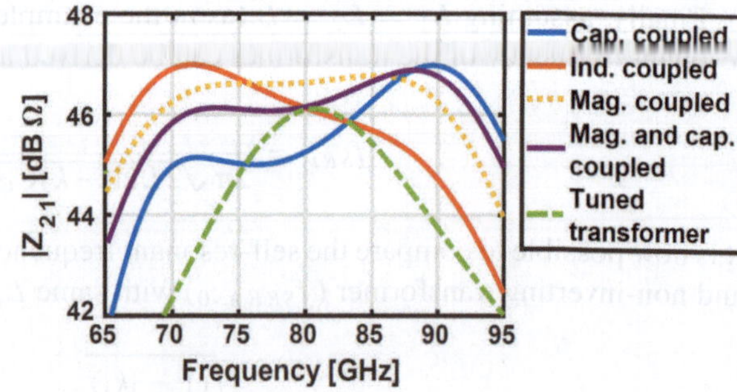

Fig. 3.12 Effect of unbalanced capacitive terminations (i.e. $C_1 \neq C_2$) on the transimpedance frequency response of different 4th order filters. ©2017 IEEE. Reprinted, with permission, from [16]

and 3.45 and the quality factor of the load is kept constant (e.g. $R_2 = 1\text{k}\Omega/2$) for fair comparison. Interestingly, the frequency response of both the magnetically coupled resonators and the single tuned transformer are not effected, except for a $10log_{10}(n)$ reduction in transimpedance gain. This is not the case for all the other 4th order filters. Capacitively and inductively coupled resonators show a balanced response if and only if $C_1 = C_2$. To solve this issue, in [13] a four-step design procedure that starts from inductively coupled resonators, applies Norton transformation, and finally derives a transformer-based filter is proposed. The end result is the same design parameters derived here in a single step form Eqs. 3.44, 3.45. Finally, the filter based on both magnetically and capacitively coupled resonators shows a frequency response in between the two, and, most importantly, the condition $C_{MC} = -C\,k_{MC}/(1 + k_{MC})$ does not result in an equalized frequency response any more.

This design example clearly shows the robustness of the proposed design techniques. Moreover, new insight is shed on these pervasive kinds of filters and simple design equations are derived.

3.3.3 Frequency Response Equalization

So far we have assumed the inductors to be lossless. When this is not the case, the frequency response of the filter shows amplitude imbalance at the two resonant peaks [13, 15, 19]. To achieve a flat frequency response without adding components or change the capacitive load, the filter in Fig. 3.7 can be redesigned by unbalancing the values of L_{M1} and L_{M2}. First, let's define the design parameter (adopting the notation in [4])

$$\xi = \frac{L_{M2}C_2}{L_{M1}C_1}. \tag{3.46}$$

When $\xi \neq 1$, the analysis of the filter response gets much more involved and the two pairs of complex poles can be written as [4, 20]

$$\omega_{L,H}^2 = \frac{1 + \xi \pm \sqrt{(1+\xi)^2 - 4\xi(1 - k_M^2)}}{2L_{M2}C_2(1 - k_M^2)}. \tag{3.47}$$

As it will be shown shortly, the design parameter ξ can be leveraged to achieve pre-emphasis. For a given value of ξ, the magnetic coupling coefficient sets the ratio ω_H/ω_L. When $\xi = 1$, Eq. 3.47 simplifies to Eqs. 3.44 and 3.45.

However, at mm-Wave frequencies the quality factor of inductors is relatively high [21, 22], therefore the required pre-emphasis needed is limited. This means that ξ close to 1 is sufficient to equalize the frequency response and Eqs. 3.44, 3.45 are still a very good approximation of ω_L, ω_H. Hence, the transformer can be designed with

$$|k_M| = \frac{\omega_H^2 - \omega_L^2}{\omega_H^2 + \omega_L^2}, \tag{3.48}$$

$$L_{M1} = \frac{1}{\omega_L^2 \, C_1 (1 + |k_M|) \sqrt{\xi}}, \tag{3.49}$$

$$L_{M2} = \frac{\sqrt{\xi}}{\omega_L^2 \, C_2 (1 + |k_M|)}. \tag{3.50}$$

The effect of ξ on the frequency response is shown in Fig. 3.13a. The remarkably simple expressions shown in Eqs. 3.49 and 3.50 shed new insights into the relation of the transformer design parameters and the filter response.

It is worth noting at this point that it is possible to equalize the filter response also by adopting other design techniques. In [15, 19] a coupling capacitor is added, resulting in the circuit in Fig. 3.8. Nonetheless, (1) capacitor losses are relatively high at mm-Wave and (2) adding capacitance to the network will result in larger ripple for the same band-pass bandwidth. Therefore, this approach is not preferable at mm-Wave. To further prove this point, the latter frequency equalization technique is compared against the proposed one in Fig. 3.13b. A pessimistic value of 10 is assumed for the quality factor of the inductors at 80 GHz. The losses are modeled with a series resistor. The lower Q-factor of the network results in lower transimpedance gain while enables to design the filter with a larger band-pass bandwidth for the same ripple. The two filters are redesigned to achieve the same $BW_{-3\,\mathrm{dB}}$ and the components values are listed in Table 3.1. As expected, adding a coupling capacitor C_{MC} results in higher in-band ripple for the same bandwidth. Moreover, when a finite quality factor $Q_{C_{MC}} = 10$ is considered, the condition $C_{MC} = -C\,k_{MC}/(1 + k_{MC})$ is not sufficient to restore a flat frequency response.

Another design procedure that starts from inductively coupled resonators (Fig. 3.6) is proposed in [13]. It is worth noting that the filter in Fig. 3.6 can not provide pre-emphasis without changing the capacitive terminations, which is not desirable. Therefore, in [13], four design parameters are introduced (i.e. d, m, l, n in [13]) and different design steps are outlined to synthesize the final transformer-based matching network without acting on the values of C_1 and C_2. When compared to the study presented here, the excellent work published in [13] lacks of simplicity and does not lead to an intuitive understanding of effect of the transformer design parameters (k_M, L_{M1} and L_{M2}) on the frequency response of the filter.

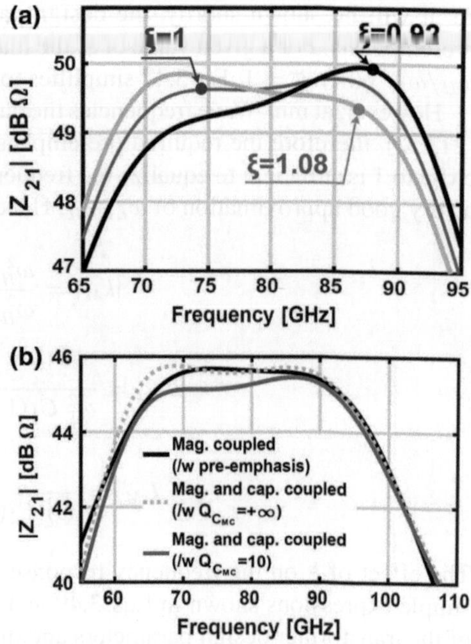

Fig. 3.13 Effect of ξ on the filter transimpedance frequency response (**a**). Comparison between different frequency equalization techniques when a limited $Q_{ind} = 10$ is considered (**b**). ©2017 IEEE. Reprinted, with permission, from [16]

Table 3.1 Component values used in the design example shown in Fig. 3.13. ©2017 IEEE. Reprinted, with permission, from [16]

$R_1\,[\Omega]$	$C_1\,[fF]$	$R_2\,[\Omega]$	$C_2\,[fF]$
400	14	1000	14
$L_{M1}\,[pH]$	$L_{M2}\,[pH]$	k_M	–
353.7	279.4	−0.362	–
$L_{MC1}\,[pH]$	$L_{MC2}\,[pH]$	k_{MC}	$C_{MC}\,[fF]$
177.3	177.3	−0.195	2.28

3.3.4 On the Parasitic Magnetic Coupling in Multistage Amplifiers

Another aspect often neglected is the effect of the parasitic magnetic coupling in multistage amplifiers. To keep the silicon area occupation as small as possible, in practice on-chip transformers are layouted physically close to each other. Therefore, some coupling is to be expected. We also expect that this effect will be exacerbated when several G_m stages are cascaded to achieve the required gain at mm-Wave.

To get insight it is instructive to refer to the schematic shown in Fig. 3.14. Here we assume to cascade three times the same transformer-based 4th order inter-stage matching network previously designed, and k_{p1} and k_{p2} are added. Figure 3.15 shows the effect of different signs of k_{p1} and k_{p2}, when $|k_{p1}| = |k_{p2}| = 0.02$. As it will be

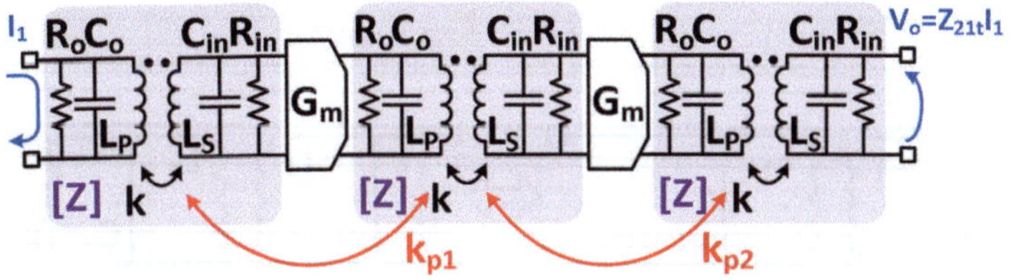

Fig. 3.14 Simplified multistage amplifier schematic with highlighted on-chip parasitic inter-stage coupling (k_{p1}, k_{p2}). ©2017 IEEE. Reprinted, with permission, from [16]

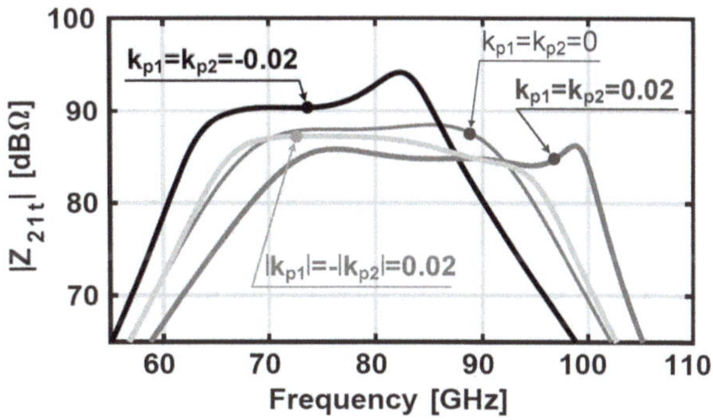

Fig. 3.15 Effect of k_{p1} and k_{p2} on the frequency response of the amplifier. ©2017 IEEE. Reprinted, with permission, from [16]

shown later, this is a reasonable assumption when transformers are layouted close to each other to save silicon area. The ideal case $k_{p1} = k_{p2} = 0$ is also reported for comparison. Clearly, the effect of the parasitic magnetic coupling is not negligible. This implies the following. (1) When a mm-Wave multistage amplifier is designed, it is important to include this effect in the EM simulations. When an extra stage is added, its matching network should be designed taking into account the effect of previous ones. (2) Ground or floating shields may be added to further limit this effect [23]. (3) The signs of k_{p1} and k_{p2} could be designed to break through the bandwidth-ripple limitation of an ideally isolated 4th order filter [14].

3.3.5 Extension to Impedance Transformation

The design example shown in Fig. 3.9 is key to develop insight into the operation of the proposed filter and absolutely relevant for on-chip inter-stage matching structures. However, for amplifiers that need to interface the input/output 50Ω environment (e.g. low-noise amplifiers LNAs and power amplifiers PA), this theory needs to be extended to realize impedance transformation.

This goal can be simply achieved by taking advantage of the properties of the transformer used to realize the 4th order filter. Figure 3.16 shows the synthesized network. Such filter shows exactly the same frequency response, while realizing a $1/n$ impedance transformation at the cost of a $1/\sqrt{n}$ reduction of the transimpedance gain

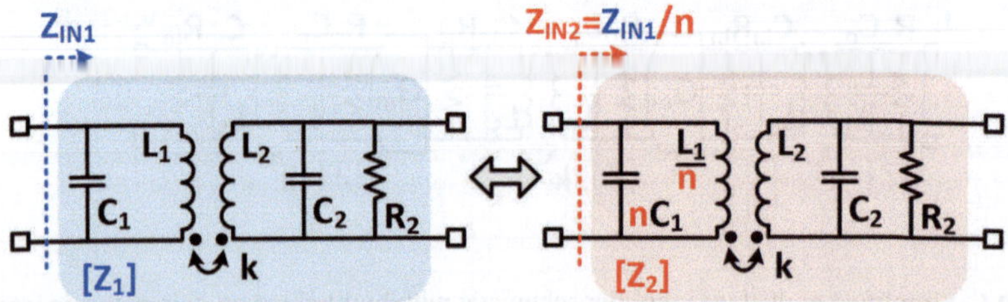

Fig. 3.16 Magnetically coupled resonator filters that realize impedance transformation

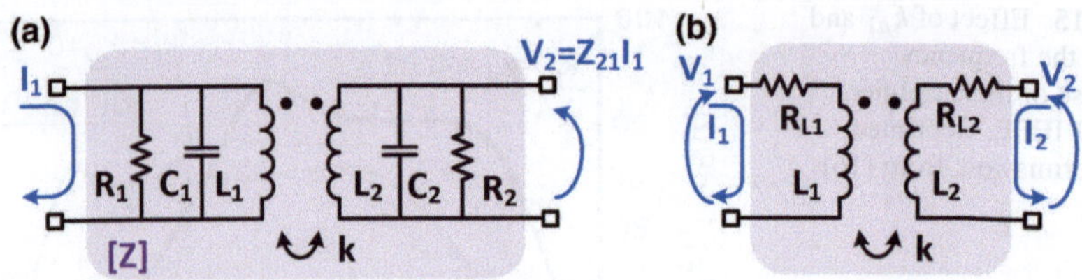

Fig. 3.17 **a** Schematic of a 4th order filter based on magnetically coupled resonator. **b** Simplified schematic of a lossy transformer

Fig. 3.18 Transimpedance frequency response of the filter as a function of k and kQ, the center frequency is kept constant to $f_o = 80\,\text{GHz}$

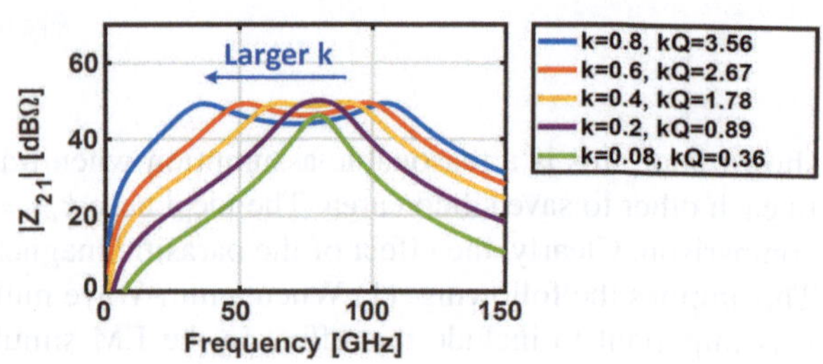

Z_{21}. The result presented here is equivalent to the one presented in [13]. However, this dissertation is remarkably simpler and does not need a much more involved Norton transformation. Moreover, the effect of the transformer design parameter on the filter response are immediately evident.

3.3.6 On the kQ Product

In the following we will further investigate the effect of the kQ product on (1) the frequency response of the filter in Fig. 3.17a, and (2) on the insertion loss of a transformer when the simplified model in Fig. 3.17b is adopted.

The kQ product is a key design parameter for the frequency response of the filter. Figure 3.18 shows the effect of the magnetic coupling coefficient on $|Z_{21}|$ of the filter based on magnetically coupled resonators (see in Fig. 3.17a). The center frequency of the filter is kept constant to

$$f_o = \frac{1}{2\pi\sqrt{C_1 L_1(1-k^2)}} = \frac{1}{2\pi\sqrt{C_2 L_2(1-k^2)}} = 80\,\text{GHz}. \tag{3.51}$$

As in the examples considered so far, $R_1 = 400\,\Omega$, $R_2 = 1\,\text{k}\,\Omega$ and $C_1 = C_2 = 14fF$, the transformer in this case is assumed lossless. As the magnetic coupling coefficient is increased the magnitude of Z_{21} increases (see Fig. 3.18) till it reaches its maximum equal to $\sqrt{R_1 R_2}/2$ at f_o when [24]

$$k = \frac{1}{Q} = \frac{1}{\sqrt{Q_1 Q_2}}, \tag{3.52}$$

where $Q_1 = \omega_o R_1 C_1$ and $Q_2 = \omega_o R_1 C_1$ are the quality factors of the filter terminations. To achieve a wideband frequency response and clearly distinguish the two resonant peaks in Fig. 3.18, the kQ product needs to be larger than 1. As the kQ product further increases a larger bandwidth is achieved at expenses of a larger in-band ripple.

When a lossy transformer is used in the filter (see Fig. 3.17b), we intuitively expect that a higher magnetic coupling factor k results in larger induced current at the secondary coil, leading to lower insertion loss. This is true as long as the quality factor of the network does not change [25, 26].

To put the foregoing discussion in perspective, let us consider a transformer with two windings L_1 and L_2. The losses of the inductors are modeled with series resistors $R_{L1} = \omega_o L_1/Q_{L1}$ and $R_{L1} = \omega_o L_2/Q_{L2}$ as in Fig. 3.17b. The efficiency and the insertion loss of this two-port network can be expressed as

$$\eta_{trasf} = \left[\frac{-1+\sqrt{1+(kQ)^2}}{kQ}\right]^2, \tag{3.53}$$

$$IL_{trasf} = 20log_{10}\left[\frac{kQ}{-1+\sqrt{1+(kQ)^2}}\right]. \tag{3.54}$$

The IL_{trasf} against the magnetic coupling coefficient for different $Q = \sqrt{Q_{L1}Q_{L2}}$ is reported in Fig. 3.19. Indeed, the lower k the higher the insertion loss. However,

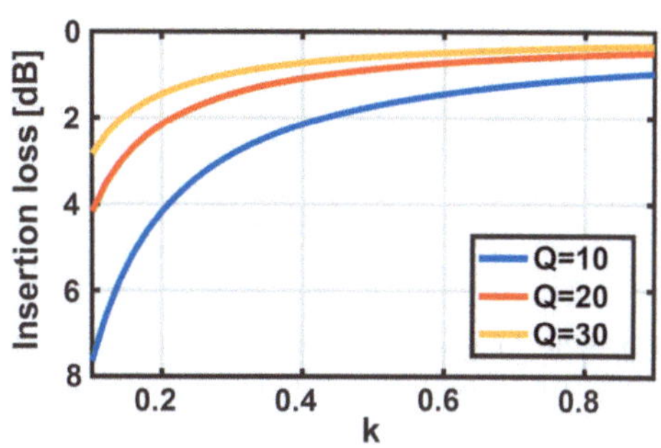

Fig. 3.19 Insertion loss of a transformer against magnetic coupling coefficient k for different Q-factors

if the quality factor is high enough, the degradation of the transformer performance is limited. The IL_{trasf} when Q is equal to 10, 20, 30 and $k = 0.4$ is respectively 1, 0.5, and 0.4 dB larger than the losses in the case of $k = 0.8$. It is worth noting that $k = 0.8$ is a typical large value when a real on-chip transformer is designed, while $k = 0.4$ is a typical value to achieve GBWEN in the inter-stage matching network of a multistage amplifier [14]. Due to technology limitations the maximum achievable Q is limited to ≈20.

0.5 dB higher insertion loss can be tolerated in a filter used for inter-stage matching. However, when the same filter is used to match the input of a low noise amplifier, it results in 0.5 dB higher noise figure. Moreover, a low-loss matching network at the output of a power amplifier is key to achieve high power added efficiency, we will return on this point in Chap. 7. Therefore, a large k is desirable for these kind of applications.

3.3.7 *Transformer-Based Power Dividers*

Power dividers are key building blocks of many receiver and power amplifier architectures. By taking advantage of the properties of transformers, magnetically coupled resonators can implement series or parallel power dividers as shown in Fig. 3.20. These 3-port networks show exactly the same frequency response of the 2-port network in Fig. 3.7 from which have been derived, except for a 3 dB lower transimpedance magnitude due to ideal power splitting.

Although these two networks are theoretically equivalent, the series power divider provides the following advantages. (1) The two required inductors at port 1 are half the value of the original L_1 and 4 times lower than the ones required in the parallel version. This leads to lower insertion loss in a practical implementation. Moreover, (2) the point of symmetry at port 1 in a series power divider is physically accessible. Making it possible to provide a symmetrical connection to the power supply and resulting in a better common mode rejection. For these reasons, series power divided are preferred in this work whenever possible.

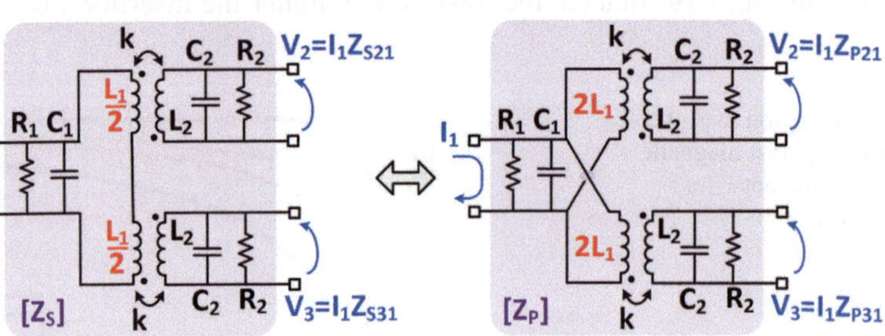

Fig. 3.20 Magnetically coupled resonator extended to realized a series power divider (shown on the left) and a parallel power divider (shown on the right)

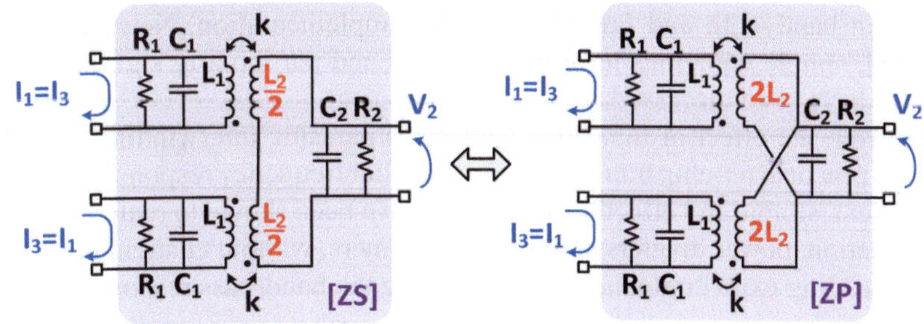

Fig. 3.21 Magnetically coupled resonator extended to realized a series power combiner (shown on the left) and a parallel power combiner (shown on the right)

3.3.8 Transformer-Based Power Combiners

To achieve the required high power levels while operating at a nominal power supply below 1V, CMOS power amplifiers largely relies on power combining techniques [21, 27–30]. 2-port magnetically coupled resonators can be extended to realize 3-port power combiners as shown in Fig. 3.21.

Although power combiners and power dividers seem to be linked by the same theory and a simple inversion of the input/output ports will do, there is a key difference that needs to be clarified. In a power divider there is only one port delivering power and the two output ports are simply loaded by a passive network. This is not the case in power combiners. The power combiners in Fig. 3.21 assume that the two input ports are driven by two currents with same magnitude and phase. When this assumption is not valid (i.e. $I_1 \neq I_3$), the analysis of the resulting network is much more involved. Even if I_1 and I_3 have the same magnitude, but different phases.

Further investigation is needed to fully understand the operation of asymmetrically driven power combiners. The study of such conditions is therefore beyond the scope of this work. However, it is worth noting that better performance at power back-off and/or lower combiner insertion losses have been shown in recent literature [29, 31–35].

3.4 Conclusion

This chapter has recalled the basics of filter design, the definition of quality factor and its effect on the filter band-pass response. Gain-bandwidth enhancement techniques have been discussed, starting from the theoretical optimum (i.e. the Bode-Fano limit), leading to the most effective topologies adopted in state-of-the-art mm-Wave design in CMOS technology.

This theoretical background has clearly pointed out that magnetically coupled resonator based on transformers stands out for their maximally-flat in-band response for a given bandwidth and favorable on-chip implementation. New insights have been shed, simple equations have been derived and particularly significant design examples have been discussed.

Moreover, the effect of the (often neglected) parasitic interwinding capacitance present in practical on-chip transformers on the filter frequency response has been discussed. Next, simple and effective techniques have been shown to realize impedance transformation, power dividers and power combiners, without changing the order of the filter, adding extra components or sacrificing the band-pass response.

References

1. B. Razavi, *Design of Analog CMOS Integrated Circuits* (McGraw-Hill Education, 2000)
2. Behzad Razavi, *RF Microelectronics*, 2nd edn. (Prentice Hall, New Jersey, 2011)
3. T.H. Lee, *The Design of CMOS Radio-Frequency Integrated Circuits* (Cambridge University Press, Cambridge, 2003)
4. A. Mazzanti, A. Bevilacqua, On the phase noise performance of transformer-based CMOS differential-pair harmonic oscillators. IEEE Trans. Circuits Syst I: Regular Pap. **62**(9), 2334–2341 (2015)
5. H.W. Bode, *Network Analysis and Feedback Amplifier Design* (D. Van Nostrand Company Inc., Princeton, 1945)
6. Robert M. Fano, Theoretical limitations on the broadband matching of arbitrary impedances. J. Frankl. Inst. **249**(1), 57–83 (1950)
7. A. Bevilacqua, A.M. Niknejad, An ultrawideband CMOS low-noise amplifier for 3.1-10.6-GHz wireless receivers. IEEE J. Solid-State Circuits **39**(12), 2259–2268 (2004)
8. D.M. Pozar, *Microwave Engineering* (Wiley, New York, 2009)
9. H. Wang, C. Sideris, A. Hajimiri, A CMOS broadband power amplifier with a transformer-based high-order output matching network. IEEE J. Solid-State Circuits **45**(12), 2709–2722 (2010)
10. W. Ye, K. Ma, K.S. Yeo, 2.5 A 2-to-6GHz Class-AB power amplifier with 28.4% PAE in 65nm CMOS supporting 256QAM, in *IEEE International Solid-State Circuits Conference - (ISSCC) Digest of Technical Papers, San Francisco, CA* (2015), pp. 1–3
11. F. Vecchi et al., A wideband receiver for multi-Gbit/s communications in 65 nm CMOS. IEEE J. Solid-State Circuits **46**(3), 551–561 (2011)
12. C.H. Li, C.N. Kuo, M.C. Kuo, A 1.2-V 5.2-mW 2030-GHz wideband receiver front-end in 0.18-μ m CMOS. IEEE Trans. Microw. Theory Tech. **60**(11), 3502–3512 (2012)
13. M. Bassi, J. Zhao, A. Bevilacqua, A. Ghilioni, A. Mazzanti, F. Svelto, A 40–67 GHz power amplifier with 13 dBm P_{SAT} and 16% PAE in 28 nm CMOS LP. IEEE J. Solid-State Circuits **50**(7), 1618–1628 (2015)
14. M. Vigilante, P. Reynaert, 20.10 A 68.1-to-96.4GHz variable-gain low-noise amplifier in 28nm CMOS, in *IEEE International Solid-State Circuits Conference (ISSCC), San Francisco, CA* (2016), pp. 360–362
15. V. Bhagavatula, T. Zhang, A.R. Suvarna, J.C. Rudell, An ultra-wideband if millimeter-wave receiver with a 20 GHz channel bandwidth using gain-equalized transformers. IEEE J. Solid-State Circuits **51**(2), 323–331 (2016)
16. M. Vigilante, P. Reynaert, On the design of wideband transformer-based fourth order matching networks for e-band receivers in 28-nm CMOS. IEEE J. Solid-State Circuits **52**(8), 2071–2082 (2017)
17. J.R. Long, Monolithic transformers for silicon RF IC design. IEEE J. Solid-State Circuits **35**(9), 1368–1382 (2000)

18. Z. Gao et al., A broadband and equivalent-circuit model for millimeter-wave on-chip M: N six-port transformers and baluns. IEEE Trans. Microw. Theory Tech. **63**(10), 3109–3121 (2015)
19. G. Li, L. Liu, Y. Tang, E. Afshari, A low-phase-noise wide-tuning-range oscillator based on resonant mode switching. IEEE J. Solid-State Circuits **47**(6), 1295–1308 (2012)
20. M. Babaie, R.B. Staszewski, A class-F CMOS oscillator. IEEE J. Solid-State Circuits **48**(12), 3120–3133 (2013)
21. D. Zhao, P. Reynaert, A 60-GHz dual-mode class AB power amplifier in 40-nm CMOS. IEEE J. Solid-State Circuits **48**(10), 2323–2337 (2013)
22. E. Mammei, E. Monaco, A. Mazzanti, F. Svelto, A 33.6-to-46.2GHz 32nm CMOS VCO with 177.5dBc, Hz minimum noise FOM using inductor splitting for tuning extension, in *IEEE International Solid-State Circuits Conference Digest of Technical Papers, San Francisco, CA* (2013), pp. 350–351
23. U. Decanis, A. Ghilioni, E. Monaco, A. Mazzanti, F. Svelto, A low-noise quadrature VCO based on magnetically coupled resonators and a wideband frequency divider at millimeter waves. IEEE J. Solid-State Circuits **46**(12), 2943–2955 (2011)
24. F. Langford-Smith, Radiotron Designer's Handbook (1941)
25. I. Aoki, S.D. Kee, D.B. Rutledge, A. Hajimiri, Distributed active transformer-a new power-combining and impedance-transformation technique. IEEE Trans. Microw. Theory Tech. **50**(1), 316–331 (2002)
26. T. Ohira, The kQ product as viewed by an analog circuit engineer. IEEE Circuits Syst. Mag. (Firstquarter) **17**(1), 27–32 (2017)
27. I. Aoki, S.D. Kee, D.B. Rutledge, A. Hajimiri, Fully integrated CMOS power amplifier design using the distributed active-transformer architecture. IEEE J. Solid-State Circuits **37**(3), 371–383 (2002)
28. P. Haldi, D. Chowdhury, P. Reynaert, G. Liu, A.M. Niknejad, A 5.8 GHz 1 V linear power amplifier using a novel on-chip transformer power combiner in standard 90 nm CMOS. IEEE J. Solid-State Circuits **43**(5), 1054–1063 (2008)
29. E. Kaymaksut, P. Reynaert, Transformer-based uneven doherty power amplifier in 90 nm CMOS for WLAN applications. IEEE J. Solid-State Circuits **47**(7), 1659–1671 (2012)
30. D. Zhao, P. Reynaert, An E-band power amplifier with broadband parallel-series power combiner in 40-nm CMOS. IEEE Trans. Microw. Theory Tech. **63**(2), 683–690 (2015)
31. W.H. Doherty, A new high efficiency power amplifier for modulated waves. Proc. Insti. Radio Eng. **24**(9), 1163–1182 (1936)
32. E. Kaymaksut, B. Franois, P. Reynaert, Analysis and optimization of transformer-based power combining for back-off efficiency enhancement. IEEE Trans. Circuits Syst. I: Regular Pap. **60**(4), 825–835 (2013)
33. E. Kaymaksut, P. Reynaert, Dual-mode CMOS doherty LTE power amplifier with symmetric hybrid transformer. IEEE J. Solid-State Circuits **50**(9), 1974–1987 (2015)
34. M. Ozen, K. Andersson, C. Fager, Symmetrical doherty power amplifier with extended efficiency range. IEEE Trans. Microw. Theory Tech. **64**(4), 1273–1284 (2016)
35. C.R. Chappidi, K. Sengupta, 20.2 a frequency-reconfigurable mm-Wave power amplifier with active-impedance synthesis in an asymmetrical non-isolated combiner, in *IEEE International Solid-State Circuits Conference (ISSCC). San Francisco, CA* (2016), pp. 344–345

Chapter 4
mm-Wave LC VCOs

The phase noise (PN) at the output of the phase locked loop (PLL) sets a fundamental limit to the maximum spectral efficiency that the whole system can achieve. As discussed in Chap. 1, the bit error rate against SNR requirements in an AWGN environment shown in Fig. 1.7 changes drastically when a practical PN profile is considered, see Fig. 1.16. Moreover, together with the tough PN requirements, a PLL should be able to synthesize the necessary LO signal over the whole band of operation.

Figure 4.1 shows the schematics of fundamental mm-Wave analog and digital PLLs. In both systems the oscillator and the first divider run at the higher frequency, limiting the PLL performance in terms of noise and power consumption. It is worth noting that the design of the analog high frequency blocks is almost identical for both analog and digital PLLs. The only difference stands in the tuning control, but the trade-offs discussed in the following applies to both voltage controlled oscillators (VCOs) and digitally controlled oscillators (DCOs).

This chapter discusses the major design challenges of mm-Wave fundamental *LC* VCOs. To achieve the noise specifications a high-Q *LC* tank is needed. Section 4.1 recalls the basics of VCOs, the linear time-variant model, the general result on PN, flicker noise upconversion and the challenges specific to mm-Wave. Section 4.2 summarizes the most popular tuning extension techniques, with a strong focus on mm-Wave applications in deep-scaled CMOS. A state-of-the-art design example of low-noise fundamental E-Band quadrature VCO is presented in Sect. 4.3.

The results of the E-Band QVCO in Sect. 4.3 have been published in European Solid State Circuits Conference (ESSCIRC 2014), IEEE Radio Frequency Integrated Circuits Symposium (RFIC 2015) and IEEE Transactions on Microwave Theory and Techniques (TMTT 2016, vol. 64, no. 4).

M. Vigilante and P. Reynaert, *5G and E-Band Communication Circuits in Deep-Scaled CMOS*, Analog Circuits and Signal Processing,
https://doi.org/10.1007/978-3-319-72646-5_4

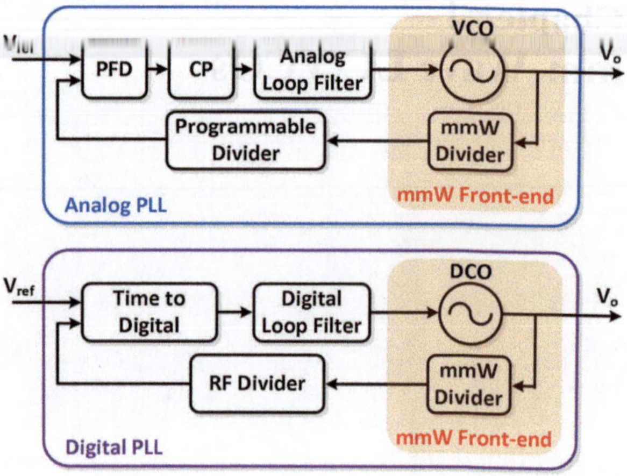

Fig. 4.1 Simplified schematic of a fundamental mm-Wave analog PLL (top) and all digital PLL (bottom)

4.1 LC VCOs Basics

4.1.1 Negative G_m Model

Figure 4.2 shows the *LC* tank highlighting the losses of a real inductor and capacitor and its equivalent parallel R_T model at the resonant frequency f_o. The quality factor of the capacitor, inductor and the resulting *RLC* tank can be expressed as

$$Q_{SC} = \frac{1}{\omega_o C R_{SC}}, \tag{4.1}$$

$$Q_{SL} = \frac{\omega_o L}{R_{SL}}, \tag{4.2}$$

$$Q_T = \frac{Im\{Z\}}{Re\{Z\}} = Q_{SC}//Q_{SL} = \omega_o R_T C = \frac{R_T}{\omega_o L}, \tag{4.3}$$

where $\omega_o = 2\pi f_o = 1/\sqrt{LC}$. By adding a negative transconductance that compensates for the tank losses R_T, it is possible to sustain a sinusoidal voltage waveform at f_o, as shown in Figs. 4.3 and 4.4.

The negative G_m model is remarkably simple and powerful. By studying the circuit in Fig. 4.3 it is possible to get insight into the operation of a complex non-linear time-variant circuit such as a VCO and draw general conclusion on the circuit-noise to phase-noise conversion [1–6].

At start-up, the circuit behaves in the small signal regime. To ensure reliable start-up conditions under PVT variations and to account for modeling inaccuracy, the slope of the transconductor I-V curve should be $g_m > 1/R_T$.

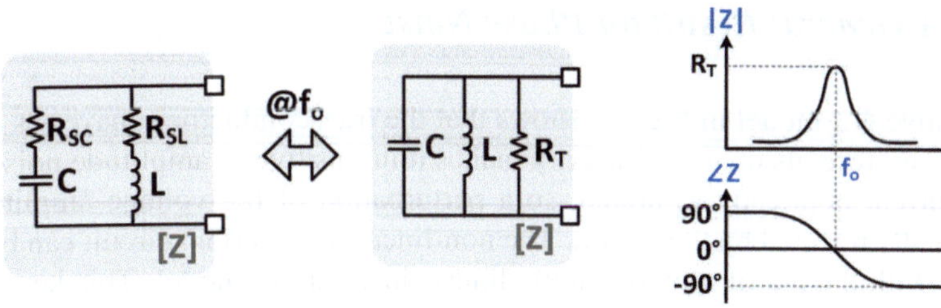

Fig. 4.2 *LC* tank loss equivalent parallel R_T model $@f_o$

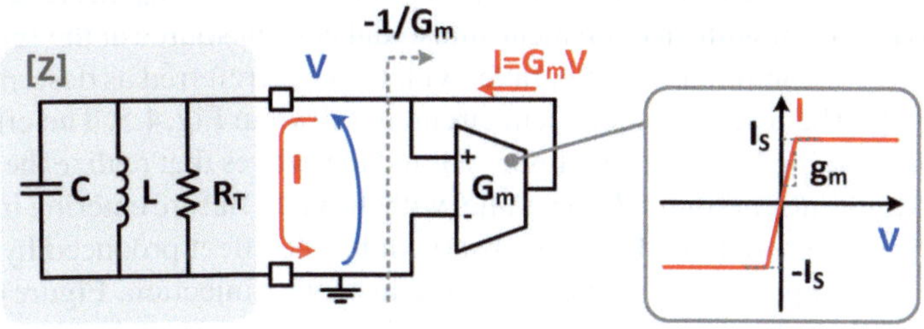

Fig. 4.3 Negative G_m model of an oscillator

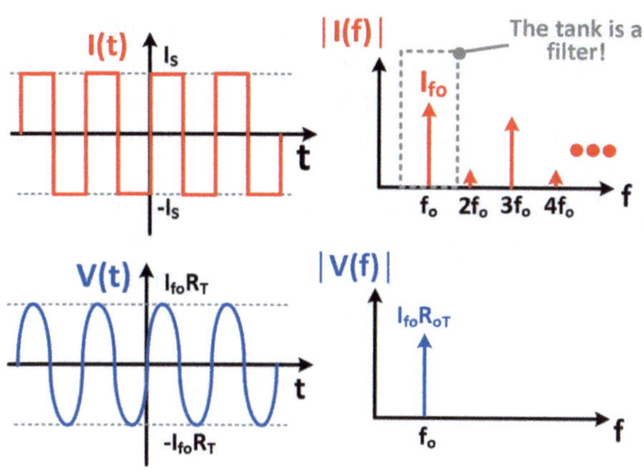

Fig. 4.4 Current and voltage waveform in the time and frequency domain

In steady state, the transconductor behaves at the first order as a hard-limiter, the tank voltage reaches its maximum and the output current saturates. Assuming that the quality factor of the *LC* tank is high enough, $Z(f)$ shows a real impedance at f_o equal to R_T and a negligible amplitude at the harmonics (Fig. 4.2). Therefore, the rich harmonics content of the current is filtered out and the tank voltage is simply

$$V = A_o \cos(2\pi f_o t) = I_{fo} R_T \cos(2\pi f_o t)\,, \tag{4.4}$$

where I_{fo} is the first harmonic of the square wave current, equal to $(4/\pi)I_s$ (Fig. 4.4).

4.1.2 A General Result on Phase Noise

The negative G_m model in Fig. 4.3 shows that the transconductor behaves as a hard limiter. Therefore, its own nonlinearity makes it insensitive to amplitude noise. The output current is already saturated, so a perturbation of the voltage amplitude is automatically rejected by the circuit. The non-linear time-variant circuit can be then analyzed with the aid of a more simple linear time-variant model. This key observation, together with the assumption of a high-Q tank that rejects the high order harmonics, is at the core of the general result on PN [3, 4].

Regardless the specific nonlinear I-V curve of the negative G_m, in steady state it is possible to replace it with an equivalent sinusoidal current source at the fundamental frequency with an amplitude I_o in parallel with the tank (referred as describing function in [1, 2]). The resulting equivalent circuit is shown in Fig. 4.5. The effect of the circuit noise coming from the tank losses R_T and the actives that realize the $-G_m$ can be modeled as a current source I_n in parallel with the tank. Before entering into further details, it is interesting to make a key observation. The effect produced by the noise current on the output voltage depends on the instant of injection. Figure 4.5 shows that a current impulse injected at the time τ_1 (i.e. when V(t) reaches its maximum) results in amplitude noise only. By the same token, a current impulse injected at the time τ_2 (i.e. when V(t) crosses the zero) affect maximally the phase and has no impact on the amplitude. Given the fact that the amplitude noise is automatically rejected, whereas a perturbation of the phase cannot be recovered and becomes phase noise, the circuit cannot be further simplified as time-invariant. This time-variant sensitivity of the circuit can be mapped in a periodic function defined as impulse sensitivity function (ISF) [2].

Although the mathematical derivation of the general result on PN is rather involved, its expression is remarkably simple. The following assumptions are made [3]. (1) The voltage waveform at the thank is sinusoidal (i.e. the tank has high Q-factor). (2) The associated ISF is sinusoidal and in quadrature with the tank voltage. (3) The active devices are always working as transconductors or off during the period. And (4) the spectral density of the transistor white current noise is proportional to its transconductance g_m.

These assumptions result in a very general theoretical best case scenario particularly relevant for a large number of practical *LC* oscillator topologies [3, 4, 6]. By referring to the schematic in Fig. 4.3 it is possible to write

$$PN(\Delta f) = 10log_{10}\left[\frac{K_B\,T}{2\,N\,Q^2\,P_{DC}\,\eta_V\,\eta_I}\left(1+\frac{\gamma}{\alpha}\right)\left(\frac{f_o}{\Delta f}\right)^2\right], \tag{4.5}$$

where K_B is the Boltzmann constant, T is the absolute temperature, N is the number of resonators, Q is the quality factor of the tank, P_{DC} is the DC power consumption, γ is the mos channel noise factor and α is a noiseless voltage gain between the tank and the transconductor stage. η_V, η_I are the voltage and current efficiencies defined as [6, 7]

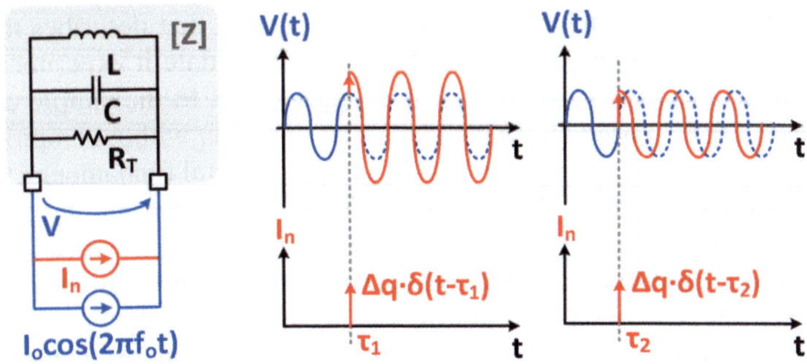

Fig. 4.5 Describing function approximation of an *LC* oscillator and its time-variant response to a noisy current impulse

Fig. 4.6 Single-sideband phase noise at the output of a free-running oscillator

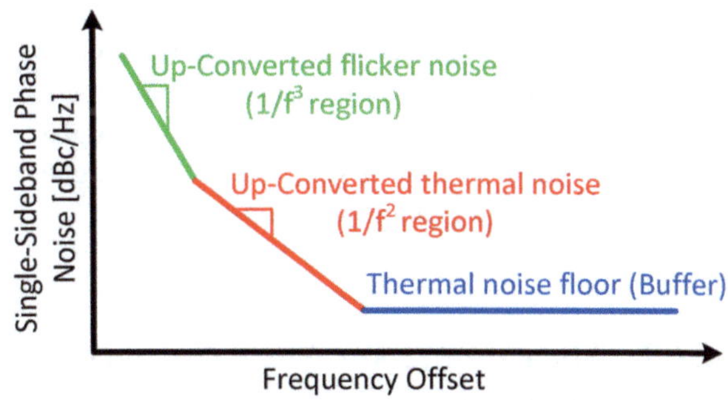

$$\eta_P = \frac{P_{RF}}{P_{DC}} = \frac{I_{RF}}{I_{DC}}\frac{V_{RF}}{V_{DC}} = \eta_V\,\eta_I, \tag{4.6}$$

where I_{RF} and V_{RF} are the current and voltage fundamental component rms values across R_T.

Figure 4.6 shows the phase noise profile at the output of a free-running oscillator. Three regions with different slopes are highlighted. The noise floor is dominated by the buffer and can be lowered to some extend at the expenses of extra DC power consumption in the LO distribution network. The oscillator can be intuitively understood as a circuit that upconverts the DC voltage to a carrier frequency f_o. Therefore, the noise closer to DC gets upconverted closer to the carrier [2, 5]. The $1/f^3$ noise is due to the flicker noise upconversion. The circuit flicker noise to phase noise conversion mechanism is particularly challenging to analyze, since it is mainly caused by second order effects and it is briefly discussed in the following subsection. Moreover, the loop bandwidth of a PLL behaves as a high-pass filter and the PN close to the carrier of the free-running oscillator is filtered out by the loop. The noise in the $1/f^2$ region is due to the circuit thermal noise upconversion and it is the most relevant for oscillator designers, since it limits the phase noise of the whole PLL far from the carrier.

In 1966 Leesons was the first to propose an equation that describes the oscillator PN in the three regions depicted in Fig. 4.6 and validate it experimentally [8]. However, it took several years and great struggle to get to the simple close form expression in Eq. 4.5 based on design parameter only (i.e. without empirical fitting factors). The general result on phase noise highlights several fundamental trade-offs. (1) The tank quality factor is key to achieve low-noise. However, any tuning technique results in a degradation of the tank Q. Therefore, oscillator with a larger tuning range (TR) are bound to have higher noise. Moreover, the ultimate limit to the tank Q for a given TR is technology itself. So this parameter is to the first order beyond the reach of the designer. (2) For a given power consumption, oscillator topologies with higher efficiency η_P result in better PN performance. (3) A noiseless voltage gain α between the tank and the input of the transconductor results in better PN performance. This could be achieved by means of a step-up transformer. By the same token, Colpitts oscillators rely on a capacitive voltage divider, realizing $\alpha < 1$ and achieving a theoretical worse PN noise for a given power consumption[1] [3, 4, 6]. (4) A straightforward way to lower the PN by a factor of N, at the fair cost of N times higher P_{DC} and extra silicon area, is to higher the number of resonators N by coupling several oscillators [10, 11]. (5) The higher the oscillation frequency f_o the worse the PN for a given power consumption.

4.1.3 More on Flicker Noise Upconversion and 2nd Order Effects

The larger the device flicker noise, the more flicker noise is upconverted. Since together with the minimum channel length at each technology node the MOS flicker noise corner gets worse, state-of-the-art oscillators with a PN flicker noise corner beyond 1 MHz have been published. Given the fact that the PLL loop bandwidth is practically limited to a maximum of $1 \sim 3$ MHz [12], techniques to minimize this phenomenon are particularly relevant.

To get insight into the circuit flicker noise to phase noise conversion a deeper understanding of 2nd order effects is needed. Equation 4.5 predicts the PN in the $1/f^2$ region, assuming that the bias tail current source in Fig. 4.7a is ideal, i.e. it is noiseless and it presents an infinite output impedance at every frequency. The design of a good current source in deep-scaled CMOS is a particularly difficult task. (1) The supply scaling is non favorable to device stacking. (2) The shorter the channel length, the lower the transistor output resistance ($R_B \propto L$) and the higher its flicker noise. On the other hand, to reduce the noise, devices with large area are preferable, exacerbating the effect of $C_B \propto W$, especially at high frequencies, see Fig. 4.7a.

[1] It is worth noting that the Colpitts oscillator can be designed to achieve lower phase noise for a given tank Q and supply voltage when compared to a class-C differential LC oscillators (up to $\approx$2 dB better). However, this comes at the expenses of a much higher current consumption and lower efficiency [9].

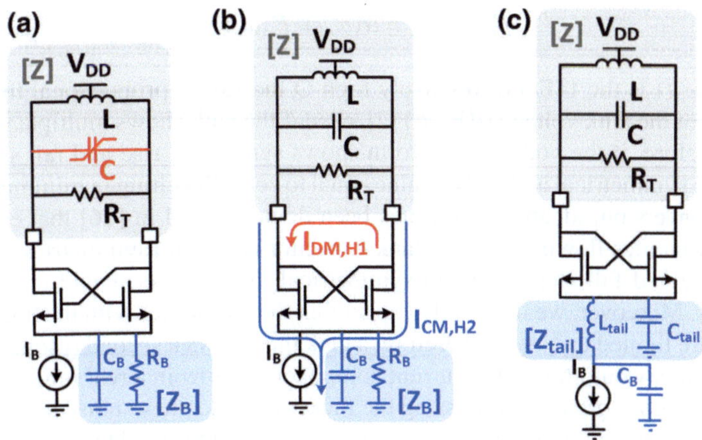

Fig. 4.7 **a** Typical cross-coupled MOS *LC* oscillator implementation with highlighted the output impedance $[Z_B]$ of a real tail current source and tank nonlinear capacitance. **b** Differential and common mode currents. **c** Improved topology with 2nd harmonic tail filter

This lead to the first flicker noise upconversion mechanism. The current source flicker noise appears in common mode and in presence of a nonlinear capacitance in the tank, results in AM-PM (i.e. phase noise) [13, 14]. This nonlinear capacitance could be somewhat small when it comes from the parasitic junction capacitance of the MOS transistors in the $-G_m$ cell or from the switches needed for tuning. But it could also be large and deliberately added to the tank to realize frequency tuning by means of varactors. This insight shows the need to lower the use of varactors to the bare minimum in order to lower the $1/f^3$ corner.

A second mechanism that seriously impacts the flicker noise to phase noise upconversion is known as Groszkowsky effect [15]. The 1st harmonic current flows in R_T, since L and C are resonating at f_o. The higher harmonics however flow in the low impedance path provided by the capacitor, perturbing the reactive energy stored in the LC tank. This results in a drift of the oscillation frequency to satisfy the resonant condition. Any variation of the harmonic content of the current at the output of the G_m devices due to $1/f$ noise, results in $1/f^3$ phase noise conversion. In theory, all the current harmonics contribute to this effect. In practice, some harmonic components are more important than others [16]. An intuitive way to comprehend this is to refer to the impulse sensitivity function (ISF) theory [2]. The DC component of the effective ISF (Γ_{eff}) is responsible for flicker noise to phase noise upconversion [2]. The flicker noise of the negative G_m stage can be modeled as a cyclostationary noise source ($i_n(t)$) in parallel with the tank as in Fig. 4.5. A cyclostationary process can be expressed as [16]

$$i_n(t) = i_{no}(\omega_o) \cdot \alpha(\omega_o t), \tag{4.7}$$

where i_{no} is a white stationary process and $\alpha(\omega_o t)$ is the noise modulating function (NMF), which is normalize, deterministic, and periodic with the maximum of 1. Once $\alpha(\omega_o t)$ is determined, the effective ISF ca be expressed as [2]

$$\Gamma_{eff}(\omega_o t) = \alpha(\omega_o t)\cdot \Gamma(\omega_o t), \tag{4.8}$$

where $\Gamma(\omega_o t)$ is the ISF. For relatively high Q the ISF is proportional to the time derivative of the tank voltage (dV/dt) [2], even if the tank shows multiple resonance [17]. Therefore, if the voltage waveform shows symmetric rise and fall slopes, the ISF will be symmetrical and its DC value equal to zero. Resulting in minimum flicker noise upconversion, ideally zero. It has been demonstrated in [16] that even-order current harmonics flowing into the tank capacitor result in asymmetries, while the voltage rise and fall slopes remain symmetrical in presence of odd-order current harmonics. Moreover, we intuitively expect that the harmonics with the highest contribution are limited to $\approx 3\omega_o$. Therefore, a tank that shows a second resonance at the 2nd harmonic, would force this current component to flow in a real impedance (and not the capacitor), greatly suppressing the flicker noise to phase noise upconversion.

A key observation that needs to be clarified is that the odd harmonics and even harmonics see two different tanks. As shown in Fig. 4.7b the odd harmonics flow in differential mode (DM) and the even harmonics flow in common mode (CM). Hence, an elegant solution to this problem is shown in Fig. 4.7c [18]. In such circuit, a second LC tank [Z_{tail}] resonating at $2f_o$ is added. This second tank appears in common mode only, therefore the operation of the oscillator at f_o is not affected. Moreover, the 2nd harmonic current sees a high impedance purely resistive [Z_{tail}] and no Groszkowsky effect can happen.

The circuit in Fig. 4.7c shows several other advantages that bring the PN of such oscillator as close as possible to the theoretical optimum of Eq. 4.5. (1) The tail capacitor C_B can now be designed large intentionally, allowing a larger parasitic capacitance coming from a real implementation of an on-chip low-noise tail current source and shunting out its noise. (2) Moreover, the extra inductor allows a larger voltage swing at the source node of the cross-coupled devices, further improving η_V. And (3), when the MOS transistors of the active core enter the triode region due to the large voltage swing, their ON resistance does not load the tank.

For all these reasons, Hegazi's VCO (2001) shows the best figure of merit (FOM) reported to date. One of the major drawback of such topology is the need for an extra on-chip inductor. This inductor has to show on the one hand high-Q and on the other has to be tunable over the oscillator TR. Several implementations that tackle these shortcomings have been recently proposed [16, 19, 20]. Interestingly, all of them are based on the observation that the tank behaves differently in DM and CM.

Another VCO circuit that raised an interesting discussion in the design community is the class-F oscillator, proposed recently in [21], and 7 years earlier with a less elegant implementation in [22]. The key idea is to leverage a 4th order tank to achieve a second resonance at $3\omega_o$. As already discussed, such topology does not provide any obvious advantage in terms of flicker noise upconversion. However, it is interesting to analyze, to understand whether or not it might be beneficial for phase noise reduction in the $1/f^2$ region, or for efficiency improvement, as it is the case for class-F power amplifiers, at least in theory. It has been proven that some noise reduction and η_V improvement when compared to the classical class-B oscillator is theoretically possible [6, 17, 23]. However, it is extremely difficult to achieve this condition in a practical implementation, while the design is definitely more involved. It is worth noting that the noise of a class-F topology could also be worse than a class-B implementation if the tank Q and equivalent parallel resistance at $3\omega_0$ are not high enough [6, 17].

Finally, it has been recently proven in a rigorous way that under certain conditions the model based on an equivalent parallel tank resistance in Fig. 4.3 that led to the general result on phase noise in Eq. 4.5 is optimistic [24]. In particular, when (1) the single-ended portion of the tank capacitance displays non-negligible losses and (2) the transconductance G_m is very large.

4.1.4 Distributed Oscillators

As discussed in Chap. 2, a T-line closed on a short circuit behaves as a frequency dependent high Q inductor. We therefore expect that when the inductor of the *LC* tank is replaced by a T-Line, it is theoretically possible to achieve higher Q at the fundamental frequency while realizing high impedance at all the odd harmonics, resulting in a clear benefit in terms of phase noise for a given power consumption. Such oscillator is referred in literature as standing wave oscillator (SWO) and its schematic is depicted in Fig. 4.8a [25]. By the same token, when the inductor is replaced with its distributed version (i.e. a T-Line), the lumped element amplifier is replaced with its distributed version (i.e. distributed amplifier) and the loop is closed in positive feedback, a new oscillator topology with superior phase noise performance can be derived [25, 26]. This circuit is referred as rotary traveling wave oscillator (RTWO) and its schematic is depicted in Fig. 4.8b.

To understand why these circuits in practice do not provide any obvious advantage in terms of phase noise performance when compared to the luped element *LC* tank counterpart, it is useful to refer to the simplified cross-section schematic in Fig. 4.8c. When a real T-line is implemented on-chip its losses are not negligible. Moreover, the losses due to the limited quality factor of the shunt capacitor have a different impact on the frequency response of the resonator when compared to the losses due to the series inductor. This is clearly shown in Fig. 4.9, where an ideal T-line

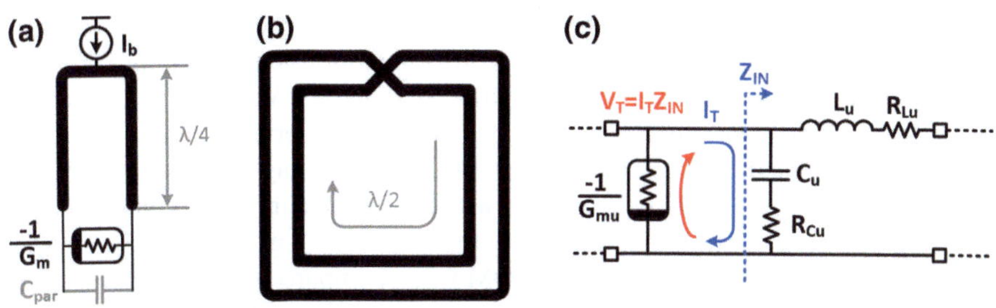

Fig. 4.8 Distributed oscillators schematics. **a** Standing wave oscillator (SWO). **b** rotary traveling wave oscillator (RTWO). **c** T-line section simplified schematic [25]

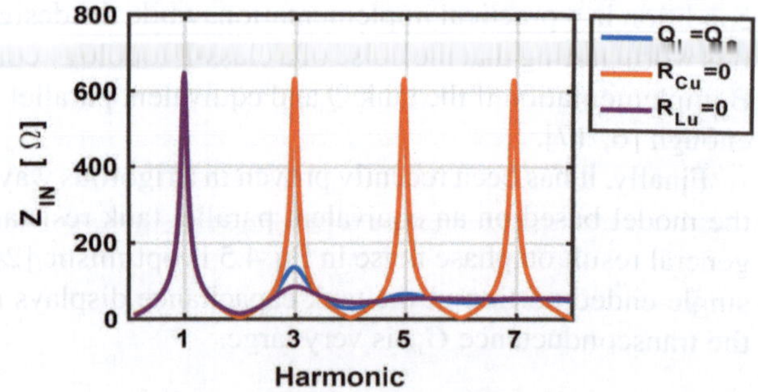

Fig. 4.9 Input impedance of a shorted T line with 50Ω characteristic impedance and finite $Q_{T-line} = 10$

is designed with a finite $Q_{T-line} = 10$ (typical value at mm-Wave). Three cases are considered, (1) when all the losses are caused by the series inductor (red line) Z_{IN} shows clear resonant peaks at the odd harmonics. (2) When the losses are all caused by the shunt capacitor (purple line) the resonant peak at $3\omega_o$ is barely visible, while at higher harmonics is not resonating anymore. (3) When the losses are equally distributed between inductor and capacitor (blue line) the resonant peak at $3\omega_o$ is heavily attenuated, posing serious doubts on the claimed advantage over the lumped element LC tank implementation. This is especially true for mm-Wave oscillators, where the tank losses are mostly capacitive, while the quality factor of inductors is relatively high. Even in the low GHz range, where capacitors perform best, when a large tuning range is needed, the quality factor of the resonator degrades so much that the voltage wave may resemble a sinusoid more than a square.

Let us now focus on the properties that distinguish these circuits from their lumped element version and render them unique. First there are several non idealities that plays a role in a practical implementation. In a SWO (Fig. 4.8a) for instance, the parasitic capacitance of the negative G_m amplifier results in an imbalance of the resonator, and in practice the T-line needs to be designed shorter than $\lambda/4$. In a RTWO (Fig. 4.8b) the inner path is shorter than the outer one, causing asymmetries. Moreover, in this case there is no physical access to a center tap, posing serious limitation on the type of amplifier that can be practically used (e.g. an inverter is the most classical solution in literature).

So far we have investigated several weak points of these circuits. However, there are also very good properties that may be extremely beneficial at system level. I.e. by tapping the T-line at different points, the RTWO permits to access ideally an infinite number of phases. Even if there is a physical limitation on the number of accessible phases, still 32, 16 and 8 have been successfully demonstrated in literature. This is a very interesting property that "comes for free" with the RTWO architecture and can be leveraged to realize a very accurate time to digital converter in an all digital PLL [27] or may be used in phased array with LO phase shifting architecture [25, 28]. Moreover, the RTWO could be combined with the SWO to include a center tap [29], allowing different possible topologies for the amplifier (e.g. NMOS CS amplifier, which can achieve lower PN for same FOM when compared to N-PMOS [23]), but also realizing an LO distribution network with several coupled oscillator with no need for power hungry buffers [25].

4.1.5 FOM and Challenges @mm-Wave

To compare different oscillator topologies and design techniques, Kinget in [30] defined a figure of merit (FOM) that normalizes the PN performance in Eq. 4.5 against power consumption and oscillation frequency as

$$FOM(\Delta f) = \frac{1}{P_{DC}\,PN(\Delta f)} \left(\frac{f_o}{\Delta f}\right)^2, \tag{4.9}$$

where P_{DC} is expressed in mW.

This FOM can be extended in an attempt to recover the tank Q degradation due to the tuning elements required to achieve a wide tuning range

$$FOM_T(\Delta f) = \frac{1}{P_{DC}\,PN(\Delta f)} \left(\frac{f_o}{\Delta f}\right)^2 \left(\frac{TR}{f_o}\right)^2. \tag{4.10}$$

This second figure of merit is a bit vague, given the fact the PN performance of any oscillator varies over the TR, especially when the TR is wide. Therefore, engineers in the literature are eager to report this FOM_T at the very best measured PN spot over the TR. Whereas, a more meaningful and interesting design challenge would be realizing an oscillator with constant PN performance over a wide tuning range.

Another figure of merit accepted from the design community is the following [31]

$$FOM_A(\Delta f) = \frac{1}{P_{DC}\,PN(\Delta f)} \left(\frac{f_o}{\Delta f}\right)^2 \frac{1}{Area}, \tag{4.11}$$

where Area is expressed in mm^2. This last FOM_A takes into account the silicon area. This is particularly relevant since as previously discussed, PN can be lowered N times without affecting the noise FOM, simply by means of N coupled oscillators [10, 11, 32]. However, a large area is normally the result of a large number of on-chip inductors, that are not desirable in practical implementation. Not only for the cost of silicon, but also because of the spurious magnetic coupling from and to other circuits.

It is worth mentioning that these figure of merits are coming from Eq. 4.5. Therefore, their validity is limited to the PN in the $1/f^2$ region.

Figure 4.10a shows a typical cross-coupled oscillator implementation, highlighting the parasitic capacitance coming from the core active devices. This fixed capacitance is in parallel with the tank and limits the oscillator tuning range

$$f_o = \frac{1}{2\pi\sqrt{L(C + C_{PAR}/2)}}. \tag{4.12}$$

Design techniques are needed to vary either the tank capacitance (C) or inductance (L) and effectively tune the oscillation frequency. In both cases, the fixed parasitic capacitance is limiting the achievable TR. This effect is particularly enhanced at

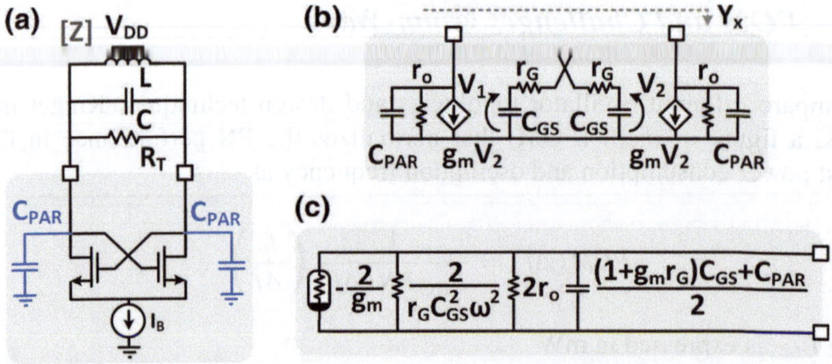

Fig. 4.10 **a** Typical cross-coupled MOS *LC* oscillator implementation with highlighted the transconductors parasitic capacitance C_{PAR}. **b** Small-signal AC model of the cross-coupled pair when the gate resistance is taken into account. **c** Equivalent parallel negative resistance model

mm-Wave, were $C_{PAR}/2 > C$! To compensate for it, more tuning elements are required, further degrading the tank Q. Therefore, larger transconductors are used, which in turns give larger fixed parasitic capacitance.

Another key observation particularly relevant for mm-Wave oscillators is that the gate resistance r_G of the transistor may significantly degrade the performance at high frequencies [33]. Figure 4.10b shows the simplified small-signal AC model of a cross-coupled pair when r_G is considered. Here C_{GD} is neglected for simplicity. The interested reader may refer to [33] for a more detailed discussion. The admittance Y_x can be expressed as

$$Re\{Y_x\} = \frac{-g_m + r_G\, C_{GS}^2\, w^2}{2(1 + r_G^2\, C_{GS}^2\, w^2)} + \frac{1}{2r_o} \approx -\frac{g_m}{2} + \frac{r_G\, C_{GS}^2\, w^2}{2} + \frac{1}{2r_o}, \tag{4.13}$$

$$Im\{Y_x\} = \frac{(1 + g_m\, r_G)C_{GS}\, w}{2(1 + r_G^2\, C_{GS}^2\, w^2)} + \frac{C_{PAR}\, w}{2} \approx \frac{(1 + g_m\, r_G)C_{GS}\, w}{2} + \frac{C_{PAR}\, w}{2}, \tag{4.14}$$

where the approximation in Eqs. 4.13 and 4.14 assume $r_G^2\, C_{GS}^2\, w^2 \ll 1$. Although this analysis is based on a small-signal model (i.e. is not valid in steady-state) it still provides insight on the start-up condition. Equation 4.13 shows that the transistor g_m needs to compensate for the losses due to its finite output impedance (r_o, proportional to the MOS channel length) and also for the losses due to the gate resistance r_G. The latter loss mechanism being proportional to ω_o^2. Moreover, Eq. 4.14 shows that the parasitic capacitance C_{CG} is magnified by a factor $1 + g_m\, r_G$.

Further, there are other effects that severely limit the performance of mm-Wave oscillators. First, due to the higher losses, devices with a large G_m are used. And, second, the quality factor of capacitors drop with frequency while inductors improve. Therefore, the fraction of tank losses due to capacitors is substantially higher than in

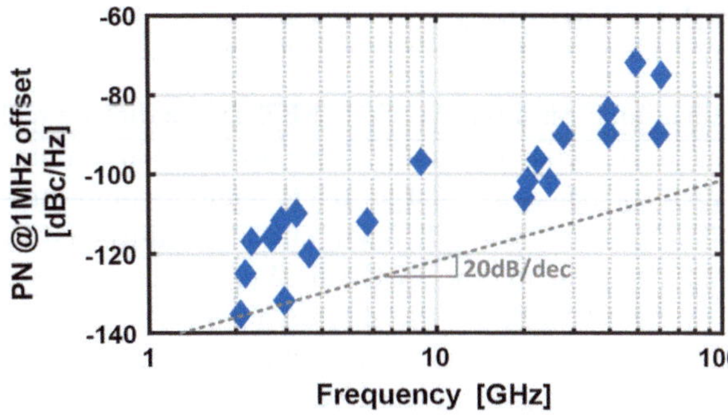

Fig. 4.11 Reported phase noise of state-of-the-art CMOS oscillators @1 MHz off-set from the carrier against oscillation frequency [34]. ©2017 John Wiley and Sons. Reprinted, with permission, from [35]

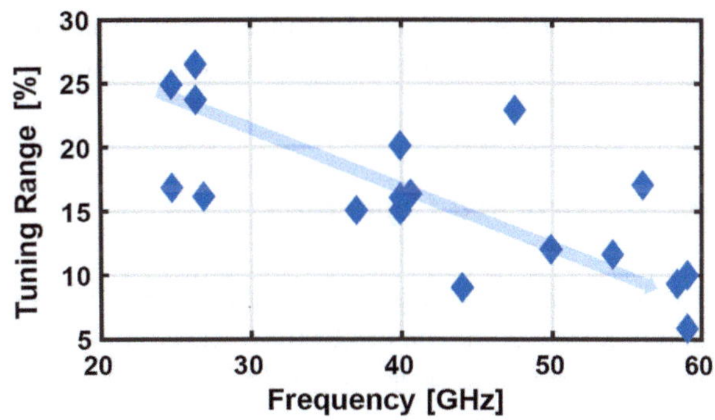

Fig. 4.12 Reported tuning range of state-of-the-art CMOS oscillators against oscillation frequency [36]. ©2017 John Wiley and Sons. Reprinted, with permission, from [35]

the low GHz range. It has been demonstrated in [24] that when these two conditions are fulfilled, the phase noise is likely to increase.

These simple design considerations lead to the empirical conclusion that reality is worse than theory. The 20 dB/dec degradation of PN performance with f_o predicted in Eq. 4.5 is indeed optimistic when state-of-the-art low-noise oscillator are considered in Fig. 4.11. Moreover, this degradation is happening together with a narrowed tuning range as clear in Fig. 4.12. This motivate the intense research efforts toward mm-Wave fundamental oscillators and frequency multipliers. Ideally, frequency multipliers with an high multiplication factor (i.e. $30 \sim 40\times$) are the solution as close as possible to the theoretical optimum. However, a practical implementation still needs a great research effort [34, 37–40].

4.2 Tuning Extension Techniques

Section 4.1 has highlighted the need for frequency tuning techniques, while Sect. 2.2 has shown the implications of scaling on the high frequency behavior of passives. In this section the insights developed so far are applied to the design of tuning circuits for mm-Wave oscillators.

4.2.1 Varactors

Accumulation-mode MOS (A-MOS) varactors are widely used in the low-GHz range to realize frequency tuning. They are normally implemented together with switched capacitor banks to enlarge to TR without compromise the AM-PM noise to phase noise conversion mechanism, ensure stability and lower the spurs level in analog PLLs by reducing the VCO gain K_{VCO} [5].

Figure 4.13 shows the shortcomings of varactors at mm-Wave. An A-MOS varactor with 200 nm channel length realized in a 28 nm CMOS process, shows a C_{MAX}/C_{min} of 3.9:1 with a quality factor lower than 10, that ranges from 9.1 to 2.6 at 80 GHz. This is due to the fact that the quality factor of capacitors is decreasing with frequency as evident in Eq. 2.14. It is therefore preferable to seek other candidates to realize wide tuning range oscillators at mm-Wave.

4.2.2 Switched Capacitors

As clear from the discussion about the effect technology scaling and BEOL metalization in Sect. 2.2, the quality factor of capacitors degrades each technology node and with the frequency of operation. Moreover, the figure of merit of switches tends to saturate in deep-scaled CMOS (as shown in Fig. 2.8). We therefore intuitively expect that switched capacitors circuit do not perform at best at mm-Wave.

Figure 4.14 shows the simplified schematic of a differential switched capacitor bank. Due to the parasitics of a real switch, the maximum achievable C_{MAX}/C_{min} is

$$\frac{C_{MAX}}{C_{min}} = \frac{C_{MOM}}{2C_{SW,OFF}}. \tag{4.15}$$

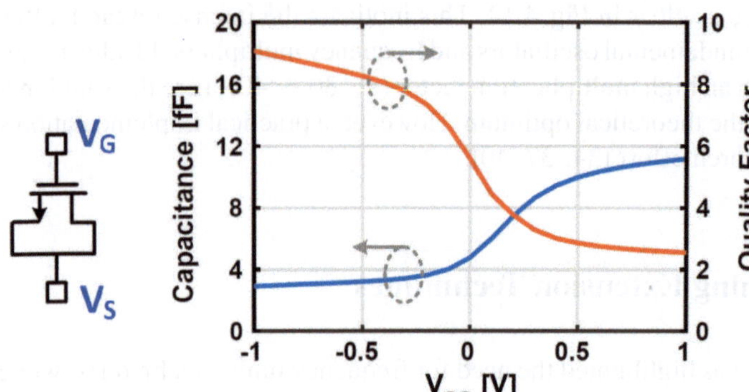

Fig. 4.13 A-MOS varactor schematic, capacitance and quality factor values against V_{GS} at 80 GHz

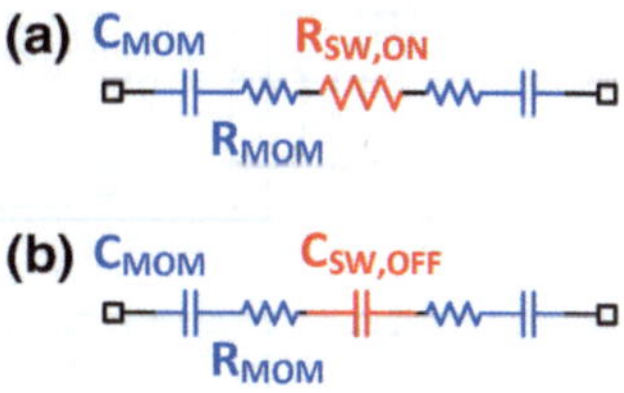

Fig. 4.14 Simplified schematic of a switched capacitor bank in **a** ON state and **b** OFF state

While the minimum quality factor of the bank is achieved in ON state and is equal to

$$Q_{min} = \frac{2}{\omega\, C_{MOM}(2R_{MOM} + R_{SW,ON})}. \tag{4.16}$$

This clearly shows the trade-off between tuning range (C_{MAX}/C_{min}) and tank quality factor (Q_{min}). It is evident that this trade-off is tighter at higher frequencies. To get deeper insight it is useful to refer to the $-G_m$ model in Fig. 4.15a. When the switch is in ON and OFF state, the respective minimum and maximum oscillation frequencies are

$$f_{min} = \frac{1}{2\pi\sqrt{L(C_{PAR} + C)}}, \tag{4.17}$$

$$f_{MAX} = \frac{1}{2\pi\sqrt{L\left(C_{PAR} + \frac{C\, C_{SW,OFF}}{C+C_{SW,OFF}}\right)}}. \tag{4.18}$$

When C_{PAR} is large it limits the maximum achievable f_{MAX}. In the following we therefore focus on alternative design techniques that are suitable for tuning extension at mm-Wave.

4.2.3 Switched Inductors

The quality factor of on-chip inductors does not significantly degrade with technology scaling. Moreover, it improves with frequency. Several techniques to switch inductors rather thank capacitors have been therefore successfully proposed at mm-Wave.

In [36], the circuit in Fig. 4.15b has been proposed. When the switch is ON it shows the same f_{min} reported in Eq. 4.17. When the switch is OFF, however, the maximum resonant frequency can be expressed as

$$f_{MAX} = \frac{1}{2\pi\sqrt{L\frac{(C+C_{PAR})C_{SW,OFF}}{C+C_{PAR}+C_{SW,OFF}}}}. \tag{4.19}$$

In this case C_{PAR} is not limiting f_{MAX} any more. $C_{SW,OFF}$ is allowed to be designed large, further improving $R_{SW,ON}$ and consequently Q_{min}.

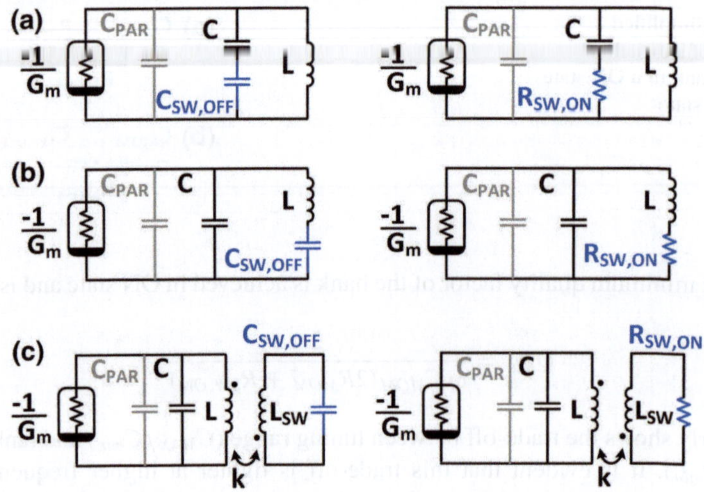

Fig. 4.15 Tuning extension techniques based on **a** switched capacitor, **b** switched inductor and **c** switched coupled inductor

Another tuning technique based on switched coupled inductors has been proposed in [41], Fig. 4.15c. This second technique is somewhat more practical when compared to the former, since it is still possible to access the center tap of the tank inductor to provide the required bias to the $-G_m$ core. When the switch is in ON state and the quality factor is high enough the equivalent tank inductance is

$$L_{eq,ON} \approx L(1-k^2). \tag{4.20}$$

whereas, when the switch is OFF the equivalent tank inductance rises to

$$L_{eq,OFF} \approx L\left(1+\frac{k^2}{\frac{\omega_{SW}^2}{\omega^2}-1}\right), \tag{4.21}$$

where $\omega_{SW}^2 = 1/(L_{SW}C_{SW,OFF})$. Once again the switch can be designed large, limiting the impact of its on resistance on the tank Q-factor. Since this time the tank shows two complex pole (this is a 4th order tank), to guarantee the oscillation at ω_o, the condition $\omega_{SW} > \omega_o$ poses an upper bound to the switch size.

4.2.4 Switched TLs

Another technique particularly effective at mm-Wave, is based on the distributed nature of transmission lines (TLs). TLs are the passive components that show the highest Q-factor and their behavior change when metal strips are added. The resulting slow-wave TL has been introduced in Sect. 2.2. By adding switches, the properties of the TL can be controlled, and the resulting circuit behaves as a distributed capacitor bank [42–44].

This technique has been effectively embedded in several circuits and demonstrates particularly fine tuning steps [45].

4.2.5 4th Order Tanks and Other Techniques

Several other tuning extension techniques have been proposed in literature. No obvious winner have been pointed out by the designers community so far. As clear from the previous discussion, there are techniques that are more favorable to technology scaling and mm-Wave applications.

Another very popular technique that exploit the two resonant frequencies of a 4th order tank have been proposed by Bevilacqua et al. in [46]. This technique in combination with switched coupled inductors resulted in a mm-Wave VCO with a record 41.1% tuning range in [47]. However, it has been proven even theoretically, that high order resonators do not provide any fundamental PN improvement when compared to the classical *LC* tank [48]. Therefore, designers still have to face the well known trade-off between TR and PN.

4.3 Design Example: A Dual-Band Transformer-Coupled QVCO in 28nm CMOS

This section presents an E-Band quadrature voltage-controlled oscillator (QVCO) implemented in 28 nm CMOS. Two fundamental oscillators are coupled by means of gate-to-drain transformers to realize accurate quadrature phases and switched coupled inductors are added for tuning extension. Closed-form expressions of the oscillation frequency and the tuning extension design parameters are derived. The time-variant nature of the circuit-noise to phase-noise of the presented topology is investigated, resulting in simple design guidelines for optimal design [49].

Based on the proposed techniques, the realized prototype is tunable over two bands of almost 5 GHz each separated in frequency, while occupying only 0.031 mm^2. The peak measured phase noise at 10 MHz offset is −117.7 dBc/Hz from a 72.7 GHz carrier and −110 dBc/Hz from a 88.2 GHz carrier and varies less than 3.5 dB within each band.

4.3.1 Proposed Transformer-Coupled Quadrature VCO

4.3.1.1 Operating Principle

Figure 4.16 shows the schematic of the proposed quadrature oscillator, where two VCOs based on the topology proposed in [50] are coupled by means of two gate-to-drain transformers. To prevent the circuit to oscillate in common mode, a resistor R_{cm} can be added on the low current path through the center tap of L_G as depicted in Fig. 4.16. The LC tank at the source node is designed with a self-resonant frequency lower than the operating frequency of the VCO, it is therefore modeled as an ideal degeneration capacitor C_2 in parallel with a RF choke inductor L_C. To gain insight into the principle of operation and simplify the following analysis, it is functional to replace the resonator with its single-ended two-port admittance parameters model as in Fig. 4.17, where $C_D = 2C_V$, $L_0 = L_G/2 = L_D/2$ and R_p accounts for the losses. By inspection, the following equations are derived:

$$Y_{11} = Y_{22} = \frac{1}{R_p} + sC_D + \frac{1}{sL_0(1 - k_{GD}^2)}, \tag{4.22}$$

$$Y_{12} = Y_{21} = \frac{k_{GD}}{sL_0(1 - k_{GD}^2)}. \tag{4.23}$$

In steady state due to the large signal operation, the non linearity of the active device and the pass-band behavior of the tank, it is possible to replace the transconductor with its describing function approximation [1]. By means of a Norton transfor-

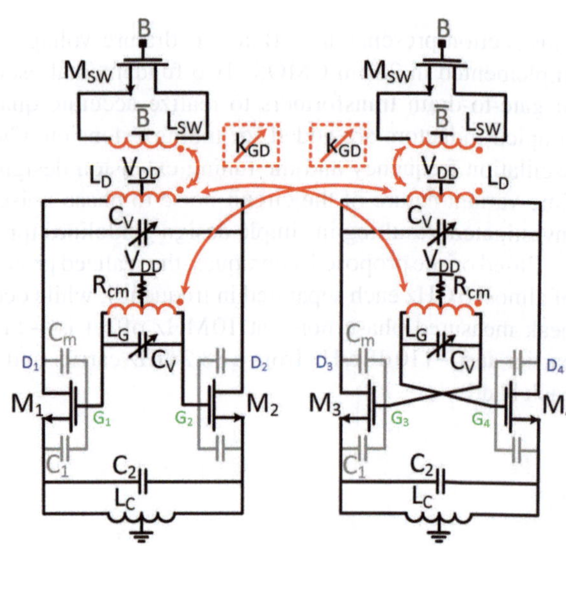

Fig. 4.16 Schematic of the proposed dual-band quadrature VCO based on gate-to-drain transformers with switched coupled inductors. ©2016 IEEE. Reprinted, with permission, from [49]

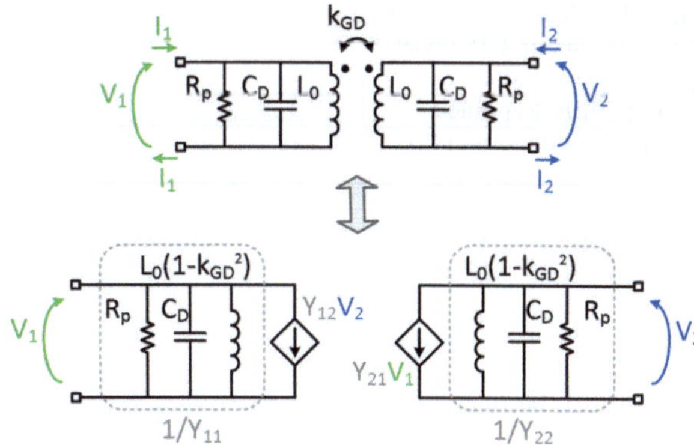

Fig. 4.17 Lumped element model of the proposed resonator and equivalent circuit based on the two-port admittance parameters matrix. ©2016 IEEE. Reprinted, with permission, from [49]

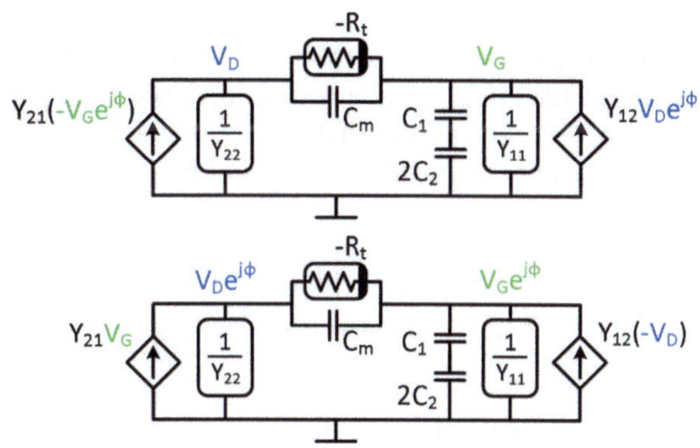

Fig. 4.18 Single-ended half circuit negative resistance AC model of the proposed QVCO. ©2016 IEEE. Reprinted, with permission, from [49]

mation and assuming differential operation (i.e. $V_{D1} = -V_{D2} = V_D$, $V_{D3} = -V_{D4} = V_De^{j\phi}$, $V_{G1} = -V_{G2} = V_G$ and $V_{G3} = -V_{G4} = V_Ge^{j\phi}$) it is possible to redraw the circuit in Fig. 4.16 as in Fig. 4.18 (see Appendix for further details). From Kirchhoff's phasor nodal equations, the following expressions are derived:

$$V_D = V_G \cdot \frac{-Y_{21}e^{j\phi} + sC_m - 1/R_t}{Y_{22} + sC_m - 1/R_t}, \tag{4.24}$$

$$V_De^{j\phi} = V_G \cdot \frac{Y_{21} + e^{j\phi}(sC_m - 1/R_t)}{Y_{22} + sC_m - 1/R_t}. \tag{4.25}$$

This set of equations is verified if and only if $\phi = \pm\pi/2$. Meaning that in presence of perfectly matched components the two oscillators are forced to operate in quadrature. Furthermore, a perfect quadrature operation is realized even if $Y_{11} \neq Y_{22}$.

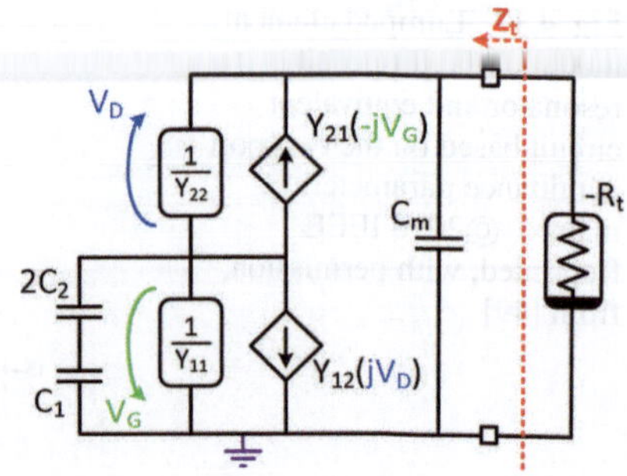

Fig. 4.19 Rearrangement of the circuit in Fig. 4.18 under quadrature operation. ©2016 IEEE. Reprinted, with permission, from [49]

4.3.1.2 Oscillation Frequency

To derive a closed-form expression for the oscillation frequency, it is effective to rearrange the circuit as a tank of impedance Z_t in parallel with an energy restoring element $-R_t$. By noting that the oscillator operates in quadrature, the circuit in Fig. 4.18 can be redrawn as shown in Fig. 4.19. By inspection, Z_t is derived as

$$Z_t(j\omega) = \frac{j\omega L_0\left[(k_{GD}^2 - 1)L_0\omega^2 + 2 - j\dfrac{2(k_{GD}^2 - 1)L_0\omega}{Rp}\right]}{Den(j\omega)}, \tag{4.26}$$

$$\begin{aligned} Den(j\omega) = &-\omega^4 L_0^2(k_{GD}^2 - 1)[(C_G + C_m)C_D + C_G C_m] + \\ &+j\frac{(k_{GD}^2 - 1)(C_D + C_G + 2C_m)L_0^2\omega^3}{R_p} + \\ +\omega^2&\left[\frac{(k_{GD}^2 - 1)L_0^2}{R_p^2} - (C_D + C_G + 2C_m)L_0\right] + j\frac{2L_0\omega}{R_p} + 1. \end{aligned} \tag{4.27}$$

Assuming an high quality factor for the resonator ($Rp \rightarrow +\infty$) and imposing the condition $Den\{Z_t\} = 0$, the two resonant frequencies of the 4th order tank are derived as

$$\omega_{1,2}^2 = \frac{2 + \alpha_1 \pm \sqrt{\alpha_2^2 + 4k_{GD}^2(\alpha_1 + \alpha_3) + 4}}{2L_0 C_m(1 - k_{GD}^2)(\alpha_1 + \alpha_3)}, \tag{4.28}$$

where $\alpha_1 = (C_D + C_G)/C_m, \alpha_2 = (C_D - C_G)/C_m, \alpha_3 = C_D C_G/C_m^2$ and $C_G = C_D + C_1 2C_2/(C_1 + 2C_2)$.

Noteworthy, this expression is similar to the one derived in [51] being the two quadrature VCOs realized around similar 4th order tanks (here C_m plays a role similar to C_C in [51]). Moreover, in principle it is possible to use the second mode of oscillation to extend further the tuning range, provided that extra circuitry is added as already proposed for other oscillator topologies [51–53]. However, given the high target frequency of operation, in this work no effort is made to take advantage of the second resonance peak, since adding extra components would lead to higher parasitic loading of the tank. Nevertheless, during the design phase it is important to ensure that the oscillator meets the Barkhausen's criteria only in one mode. Condition easily achieved in a practical design at mm-Wave frequencies when the magnetic coupling k_{GD} is designed large enough, so that $\omega_2 \gg \omega_1$ and the transconductors are not able to compensate for the tank losses in the second mode.

4.3.1.3 Tuning Extension

In *LC* oscillators higher tuning range comes at the cost of lower spectral purity for a given power consumption. This trade-off is exacerbated at mm-Wave, where the impact of parasitics is larger and the quality factor of the tank is limited by the Q of capacitors and varactors rather than the one of inductors. For these reasons, several recent research works have been focused on alternative tuning extension techniques [36, 41, 47].

From Eq. 4.28 it is clear that the oscillation frequency is highly sensitive to the magnetic coupling coefficient k_{GD} and the inductance value L_0. In this work, a third winding L_{SW} terminated on a switch M_{SW} is coupled to the gate-to-drain transformer to effectively change both k_{GD} and L_0, as depicted in Fig. 4.20. Intuitively, when the switch is turned ON, the current induced in L_{SW} through k_2 and k_3 finds a low impedance path. Whereas, when M_{SW} is in OFF state, L_{SW} is terminated on an infinite impedance and ideally no current is flowing. To gain deeper insight into the operation of the proposed transformer with switched coupled inductor and derive design guidelines, it is useful to refer to its three-port impedance parameter model shown in Fig. 4.21. By inspection the following expression are derived

$$Z_{11} = R_G + sL_G, \tag{4.29}$$

$$Z_{22} = R_D + sL_D, \tag{4.30}$$

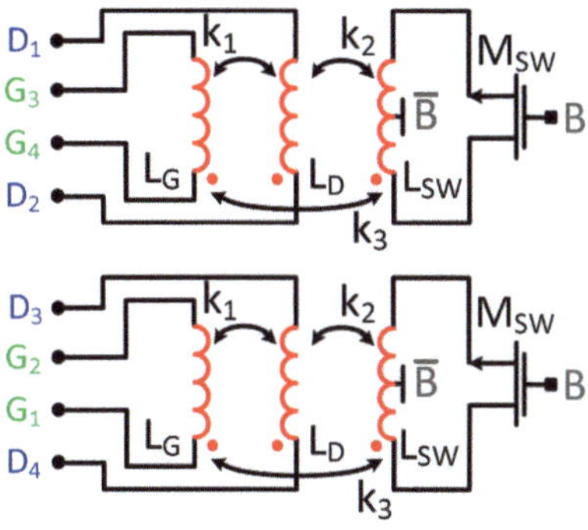

Fig. 4.20 Simplified lumped element model of the gate-to-drain transformers with switched coupled inductors. ©2016 IEEE. Reprinted, with permission, from [49]

$$Z_{33} = R_{SW} + sL_{SW}, \tag{4.31}$$

$$Z_{12} = Z_{21} = sk_1\sqrt{L_G L_D}, \tag{4.32}$$

$$Z_{23} = Z_{32} = sk_2\sqrt{L_D L_{SW}}, \tag{4.33}$$

$$Z_{13} = Z_{31} = sk_3\sqrt{L_G L_{SW}}, \tag{4.34}$$

where the losses of each inductor (L_G, L_D and L_{SW}) in Fig. 4.20 are modeled by a series resistor (of value R_G, R_D and R_{SW} respectively).

Since the third port is terminated on an impedance Z_{SW}, it is possible to derive the equivalent two-port network as depicted in Fig. 4.21, provided that

$$\widetilde{Z}_{11} = Z_{11} - \frac{Z_{13}Z_{31}}{Z_{SW} + Z_{33}}, \tag{4.35}$$

$$\widetilde{Z}_{12} = Z_{12} - \frac{Z_{13}Z_{32}}{Z_{SW} + Z_{33}}, \tag{4.36}$$

where for sake of space only the expressions for the impedance and the transimpedance of the first winding are reported.

It is now possible to derive approximated equations to describe an equivalent two-port transformer. When M_{SW} is in ON state $Z_{SW} \approx R_{ON}$ and assuming $R_{SW} + R_{ON} \ll \omega L_{SW}$, the equivalent series resistance and self-inductance of the primary winding ($R_{G,ON}$, $L_{G,ON}$) and the equivalent magnetic coupling $k_{GD,ON}$ are

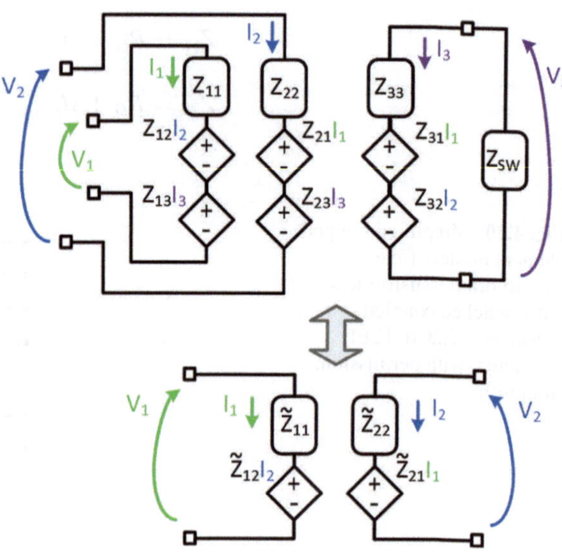

Fig. 4.21 Transformer with switched coupled inductor three-port impedance parameters model (top) and its two-port equivalent circuit (bottom). ©2016 IEEE. Reprinted, with permission, from [49]

$$R_{G,ON} \approx R_G + k_3^2 \frac{L_G}{L_{SW}}(R_{SW} + R_{ON}), \tag{4.37}$$

$$L_{G,ON} \approx L_G - L_G k_3^2, \tag{4.38}$$

$$k_{GD,ON} \approx \frac{k_1 - k_2 k_3}{\sqrt{(1-k_2^2)(1-k_3^2)}}. \tag{4.39}$$

When M_{SW} is in OFF state $Z_{SW} \approx 1/(sC_{OFF})$ and assuming $R_{SW}^2 \ll (\omega L_{SW} - 1/(\omega C_{OFF}))^2$, the following expressions are derived

$$R_{G,OFF} \approx R_G + \frac{k_3^2}{(\omega_{SW}^2/\omega^2 - 1)^2} \frac{L_G}{L_{SW}} R_{SW}, \tag{4.40}$$

$$L_{G,OFF} \approx L_G + L_G \frac{k_3^2}{\omega_{SW}^2/\omega^2 - 1}, \tag{4.41}$$

$$k_{GD,OFF} \approx \frac{k_1 + \dfrac{k_2 k_3}{(\omega_{SW}^2/\omega^2 - 1)}}{\sqrt{\left[1 + \dfrac{k_2^2}{\omega_{SW}^2/\omega^2 - 1}\right]\left[1 + \dfrac{k_3^2}{\omega_{SW}^2/\omega^2 - 1}\right]}}, \tag{4.42}$$

where $\omega_{SW}^2 = 1/(L_{SW}C_{OFF})$ is the self-resonant frequency of the third winding when terminated on the OFF capacitance of M_{SW}.

From Eqs. 4.37–4.42 several design considerations can be made. First, the ON resistance of M_{SW} severely increases the transformer losses and should be designed accordingly low enough. In addition, the current induced in the third winding effectively reduces both $L_{G,ON}$ and $k_{GD,ON}$ through k_2 and k_3.

Another critical observation deals with the design of ω_{SW} when M_{SW} is in OFF state. To ensure a single solution of Eq. 4.41, ω_{SW} should be higher than the oscillation frequency ω_o of the VCO, imposing an upper bound for the value of C_{OFF} [41, 47]. Moreover, in a practical design, the condition $\omega_{SW} \gg \omega_o$ is not verified. Meaning that $R_{G,OFF}$, $L_{G,OFF}$ and $k_{GD,OFF}$ increase with frequency and the change is sharper when ω_o approaches ω_{SW}, demanding for a careful co-design of M_{SW}, L_{SW}, k_2 and k_3.

4.3.1.4 Phase Noise

The first step towards a design for minimum phase noise is to quantify the effect of two key design parameters (i.e. the equivalent magnetic coupling k_{GD} and the degeneration capacitance C_2) on the operation of the proposed quadrature oscillator.

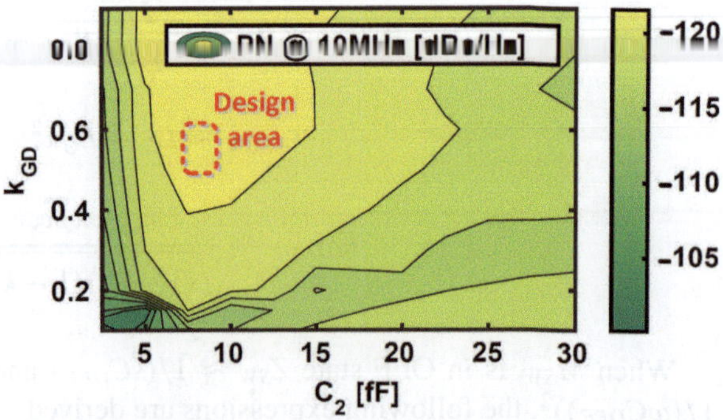

Fig. 4.22 Simulated phase noise at 10 MHz offset from an 80 GHz carrier versus k_{GD} and C_2.

Figure 4.22 shows the simulated phase noise at 10 MHz offset from an 80 GHz carrier as a function of k_{GD} and C_2. The experiments were performed adopting ideal lumped element components for passive devices where $C_V = 10\,\text{fF}$, L_0 is adjusted to keep the oscillation frequency equal to 80 GHz for fair comparison and the losses are modeled with a shunt resistor assuming a quality factor equal to 4. The transconductors (W/L of 40 μm/28 nm) were post-layout parasitic extracted to account for C_m and C_1 (about 10 and 20 fF respectively). R_{cm} is set to 100Ω to prevent oscillation in common mode. Clearly, the phase noise shows a weak dependency from k_{GD}, meaning that the proposed tuning extension technique can be effectively applied provided that the quality factor of the resonator is kept constant when M_{SW} is in ON and OFF state. Moreover, the value of C_2 can be optimized for PN.

To get deeper insight into the circuit-noise to phase-noise conversion mechanism of the proposed topology it is useful to adopt the linear time-variant approach proposed by Hajimiri and Lee in [2]. By noting that in steady state the oscillation amplitude is limited by the compressing behavior of the transconductors, the amplitude noise is neglected and the phase noise at an angular frequency offset $\Delta\omega$ from a ω_o carrier can be expressed as

$$L(\Delta\omega) = 10log_{10}\left(\frac{\sum_i N_{L,i}}{2q_{max}^2(\Delta\omega)^2}\right), \tag{4.43}$$

where q_{max} is the maximum charge displacement across the tank capacitor and $N_{L,i}$ is the effective noise power of the ith current noise source, defined as

$$N_{L,i} = \frac{1}{T_o}\int_0^{T_o} \Gamma_i^2(t)\,\overline{i_{n,i}^2(t)}\,dt, \tag{4.44}$$

where $\Gamma_i(t)$ is the impulse sensitivity function (ISF), dimensionless function of time periodic in $T_o = 2\pi/\omega_o$ and $\overline{i_{n,i}^2}$ is the power spectral density of the current noise produced by the ith devices.

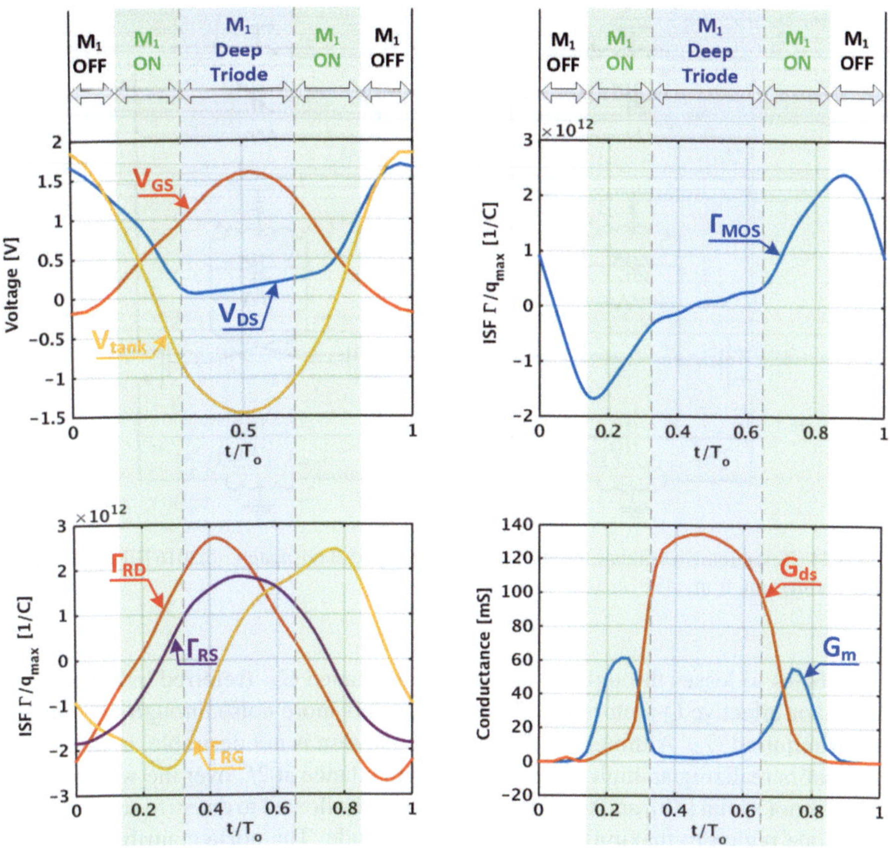

Fig. 4.23 Noise to phase noise conversion mechanism of the proposed oscillator. **a** Voltage waveforms and M_1 operation region. **b** Impulse sensitivity function of M_1 channel noise. **c** Impulse sensitivity function at different nodes. **d** Instantaneous transconductance (G_m) and channel conductance (G_{ds}) of M_1. ©2016 IEEE. Reprinted, with permission, from [49]

The ratio $\Gamma_i(t)/q_{max}$ is accurately estimated at different oscillator nodes by means of the periodic transfer function analysis (PXF) as proposed in [54] and reported in Fig. 4.23c. In these experiments $k_{GD} = 0.5$ and the optimum value of C_2 in Fig. 4.22 is selected. Most importantly, Fig. 4.23c shows that $\Gamma^2_{RS,rms} \approx 0.65\ \Gamma^2_{RG,rms} \approx 0.54\ \Gamma^2_{RD,rms}$, meaning that the source node is the less sensitive to noise and the drain node is the most critical one. To optimize the phase noise performance is therefore important to design the transformer accordingly (i.e. the quality factor of the secondary winding Q_D should be maximized). Moreover, the output should be probed at the source to minimize the loading effect of the following stage.

Figure 4.23a shows the voltage waveforms across M_1 highlighting its regime of operation. Due to the large voltage amplitude the transistor operates in all three regions (i.e. saturation, triode and OFF), meaning that the active device (1)

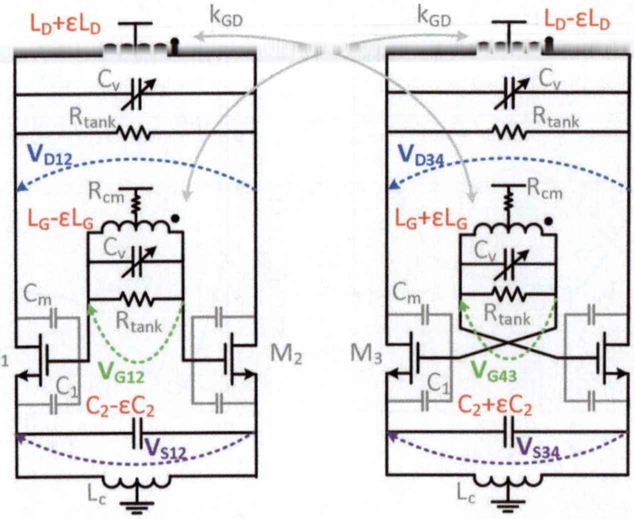

Fig. 4.24 Test circuit to evaluate the phase error in presence of mismatch. ©2016 IEEE. Reprinted, with permission, from [49]

contributes to losses through the channel conductance G_{ds} (referred in literature as loaded or effective Q-factor [4, 55]) and (2) injects more noise through G_{ds} and the larger required G_m. Although in theory this condition is not desirable, in mm-Wave oscillators realizing an high common mode impedance at $2f_o$ over the whole tuning range is not trivial [56], and the transconductors are allowed to enter (to some extend) the triode region to maximize the voltage amplitude. The noise contribution of the transistor can be expressed as [4]

$$\overline{i_{n,MOS}^2}(t) = 4K_B T(\gamma G_m(t) + G_{ds}(t)), \tag{4.45}$$

where K_B is the Boltzmann's constant, T absolute temperature and γ the transistor channel noise factor.

Noteworthy, Fig. 4.23b, d show that when the transistor noise contribution is at its maximum, the associated Γ_{MOS} is close to 0 yielding a negligible associated effective noise power $N_{L,MOS} \approx 0$ during this lap of time.

4.3.1.5 Phase Error

In presence of mismatches in the circuit Eqs. 4.24 and 4.25 are not valid anymore and the two oscillators depart from quadrature. This result in amplitude imbalance and phase error. The focus of this section is the phase error, since in a practical system the LO signals are normally fed to hard limiting buffers and an I/Q mixer that is almost insensitive to small amplitude imbalance (provided that the signal amplitude

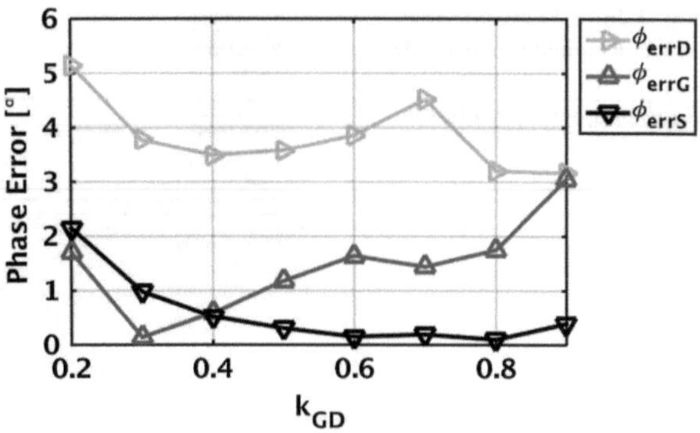

Fig. 4.25 Simulated impact of 2ε=2% mismatch among passives on the phase error at different nodes against k_{GD}. ©2016 IEEE. Reprinted, with permission, from [49]

is large) [57, 58]. Deriving elegant closed-form expressions for the phase error at different tank nodes under the presence of mismatches for this QVCO topology is not trivial. Moreover, the simplified linear model in Fig. 4.18 when extended to describe mismatch due to the tank, would not account for circuit nonlinearity that may play a significant role (as is the case for QVCO that employs passive nonlinear couplers [58]).

To address this problem, it is functional to refer to the schematic depicted in Fig. 4.24. To gain a deeper understanding, a mismatch of $2\varepsilon = 2\%$ is imposed among passives and the phase error is evaluated at different nodes. Figure 4.25 shows that when the equivalent magnetic coupling is $k_{GD} > 0.3$ the phase error at the source node is always $\phi_{errS} \leq 1°$ and most importantly when $k_{GD} > 0.4$ the condition $\phi_{errS} \leq \phi_{errG} \leq \phi_{errD}$ is achieved.

From the analysis above, we can draw two very important and perhaps unexpected conclusions, (1) probing the signal at the source leads to the same choice of optimal design parameters for both minimum phase noise (as in Fig. 4.22) and minimum phase error (Fig. 4.25), and (2) the loading effect of the buffer is minimal at this node (as clear from Fig. 4.23c and already discussed in Sect. 4.3.1).

4.3.2 *Design Considerations at mm-Wave and Circuit Implementation*

Thanks to the aggressive scaling of the gate length, nowadays mm-Wave circuits can enjoy active devices with a f_t as high as 300 GHz in technology nodes as 28 nm CMOS [59]. However, when these transconductors with high intrinsic performance are used in *LC* oscillators the effect of the parasitics due to the layout interconnects severely limits the improvement in terms of effective f_t and yields large fixed capacitance making the tuning range versus phase noise trade-off tighter [60]. Furthermore, in deep-scaled CMOS processes (1) low level metals get thinner and closer to the substrate, reducing the quality factor of metal-oxide-metal (MOM) capacitors, (2) high level metals get closer to the substrate, lowering the achievable Q of inductors and (3)

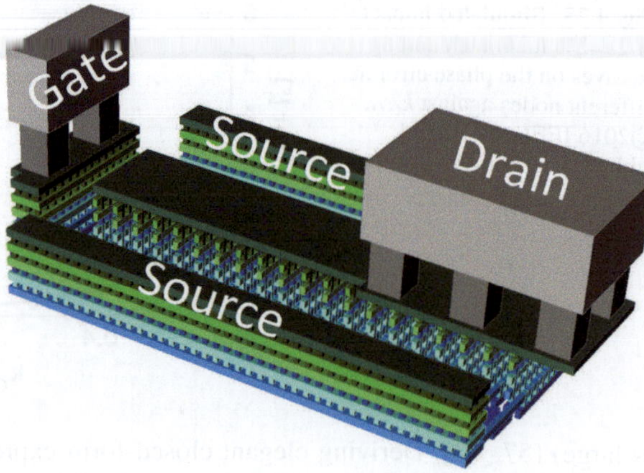

Fig. 4.26 3-D layout view of the designed transistor M_1 40 μm/28 nm. ©2016 IEEE. Reprinted, with permission, from [49]

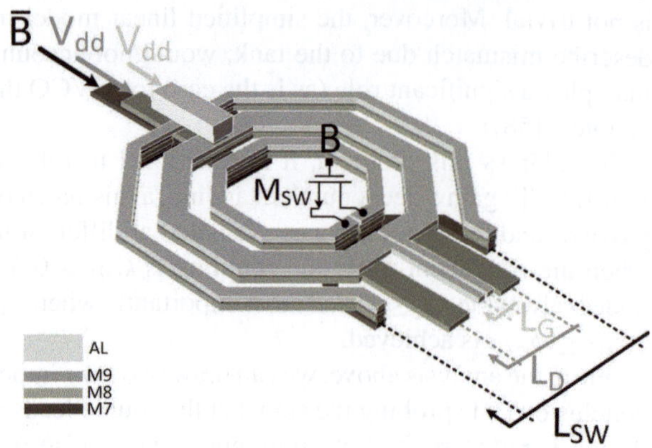

Fig. 4.27 3-D layout view of the designed transformer with switched coupled inductor. ©2016 IEEE. Reprinted, with permission, from [49]

the design rule check (DRC) imposes ever increasing minimum metal density to be fulfilled in tighter area windows, limiting even further the maximum achievable Q-factor and self-resonant frequency of on-chip inductors [36, 61]. A number of design techniques are discussed in this section to tackle the aforementioned challenges for mm-Wave *LC* oscillators.

As a first step, we focus on the design and layout of the active core. The parasitic gate-to-drain capacitance plays an important role in the design of any *LC* oscillator, lowering the oscillation frequency and limiting the tuning range [5], and the presented topology is no exception. As a matter of fact, C_m shown in Fig. 4.16 appears single-ended, lowering the oscillation frequency as clear from Eq. 4.28. This capacitance is kept minimum by adopting the transistor layout presented in [60] and shown in Fig. 4.26. Moreover, thanks to this layout, it is now possible to access directly the gate and drain of the transistor in higher metal, limiting the losses due to interconnections to the tank. The source node is accessed at both sides, minimizing the critical gain reduction due to the connection to this net and simplifying the routing to C_2 and L_C shown in Fig. 4.16. In this design, the transistors are oversized to 40 μm/28 nm to account for possible model inaccuracy.

Another key aspect of any mm-Wave *LC* oscillator is the design and layout of the tank. Figure 4.27 shows the 3-D view of the layout of the proposed gate-to-drain transformer. A relatively high value of the magnetic coupling coefficient between the primary and secondary windings k_1 (see Fig. 4.20) is desirable, so that it becomes the dominant factor in the expressions of equivalent magnetic coupling k_{GD} in ON and OFF state Eqs. 4.39, 4.42. Since the value of the required self-inductances is relatively low, to maximize the magnetic coupling L_G and L_D are realized as an overlay transformer in metal 9 and 8 respectively (see Fig. 4.27), with a metal width of 4 μm and an outer diameter of 37.8 μm.

The switched coupled inductor L_{SW} in Fig. 4.20 is realized with an inner coil and an outer coil in both metal 8 and 9 with 2 μm width, connected together in metal 7. The inner and outer spacing of L_{SW} from the primary and secondary windings (i.e. L_G and L_D) are 2.9 and 3.5 μm respectively (as shown in Fig. 4.27). As discussed in Sect. 4.3.1 the value of R_{ON} and C_{OFF} of the switch M_{SW} proves critical. Since at mm-Wave the inductor Q-factor is relatively high, the value of R_{ON} will dominate the losses of the transformer in the ON state, as predicted from Eq. 4.37. M_{SW} is therefore designed large, with a W/L of (39 × 3) μm/28 nm. To further optimize the switch Figure of Merit (FOM$_{SW}$ $= R_{ON}C_{OFF}$), the source and drain connections of M_{SW} are layouted with a tapered via stack to minimize C_{OFF}. Figure 4.28 shows the proposed gate-to-drain transformer with switched coupled inductor simulated parameters when the switch is in ON and OFF state. From electromagnetic simulation the equivalent magnetic coupling coefficient (Fig. 4.28a), self-inductances (Fig. 4.28b) and quality factors (Fig. 4.28c) of primary and secondary windings of the transformer when M_{SW} is OFF (ON) are $k_{GD} = 0.59$ (0.5), $L_G = 100$ pH (82 pH), $L_D = 92$ pH (75 pH), $Q_G = 8$ (6), $Q_D = 13$ (9) at 73.5 GHz (83.5 GHz). It is worth to mention that, as clear from the discussion about circuit-noise to phase-noise conversion in Sect. 4.3.1 and shown in Fig. 4.28c, the winding with the higher quality factor is reserved to L_D. To compensate for the degradation of the tank Q in the higher band (i.e. when M_{SW} is in ON state), in this work the value of the degeneration capacitance C_2 is designed for optimal phase noise in this mode of operation, aiming at an uniform noise FOM over the whole tuning range. To further tune the oscillator continuously within the two bands, two binary-weighted digitally controlled MOM capacitors and an accumulation-mode MOS (A-MOS) varactor are added to the tank.

To minimize the flicker noise to phase noise upconversion, a voltage-biased topology is adopted in this design. Removing the current control is a critical choice, common to several state-of-the-art low-noise mm-Wave *LC* oscillators (such as [36, 40, 56, 62]). In fact, the lack of ideal current sources is exacerbated at high frequencies by the larger effect of the parasitic capacitance to the substrate.

Figure 4.29 shows the block diagram of the realized chip. For measurement purpose, two buffers and an I/Q double-balanced mixer are also implemented on-chip. The buffers are realized with pseudo-differential neutralized common source amplifiers, providing high input-output isolation, driving the 50 Ω measurement equipment directly at mm-Wave and controlling the on-chip mixer. The latter is based on

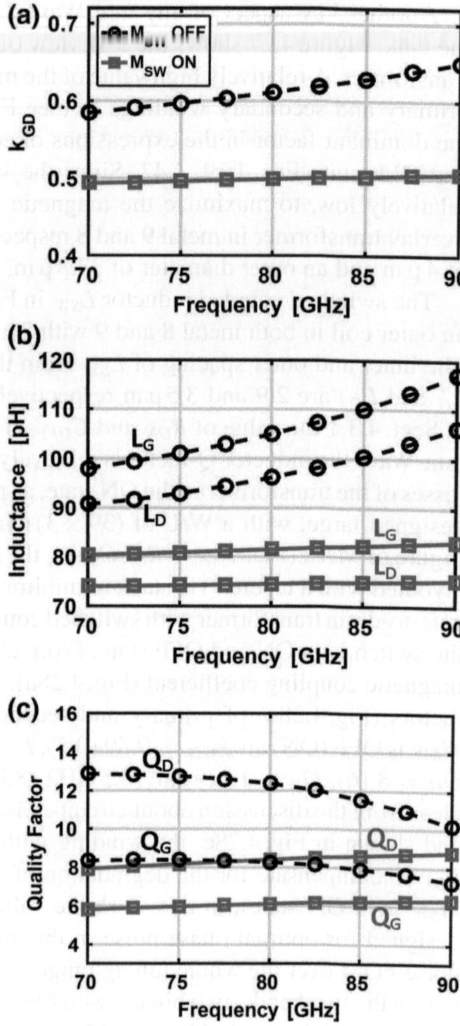

Fig. 4.28 Simulated characteristics of the gate-to-drain transformer with switched coupled inductor when M_{SW} is ON (solid grey line) and OFF (dashed black line) against frequency. **a** Equivalent magnetic coupling k_{GD}. **b** Self-inductance of the primary and secondary winding (L_G and L_D). **c** Quality factor of the primary and secondary winding (Q_G and Q_D). ©2016 IEEE. Reprinted, with permission, from [49]

a Gilbert cell, allowing the downconversion of the high frequency on-chip quadrature signals to an intermediate frequency, instrumental to measure I/Q amplitude and phase imbalance.

4.3.3 Measurement Results

Figure 4.30 shows the die micrograph of the quadrature VCO prototype fabricated in 28 nm bulk CMOS technology with no RF thick metal option. It occupies an

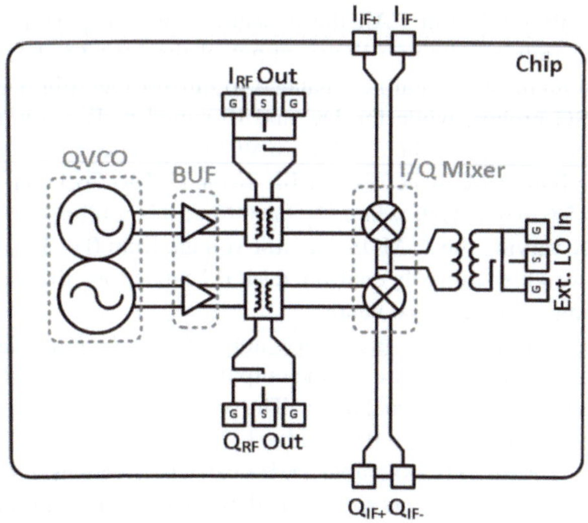

Fig. 4.29 Block diagram of the realized test chip. ©2016 IEEE. Reprinted, with permission, from [49]

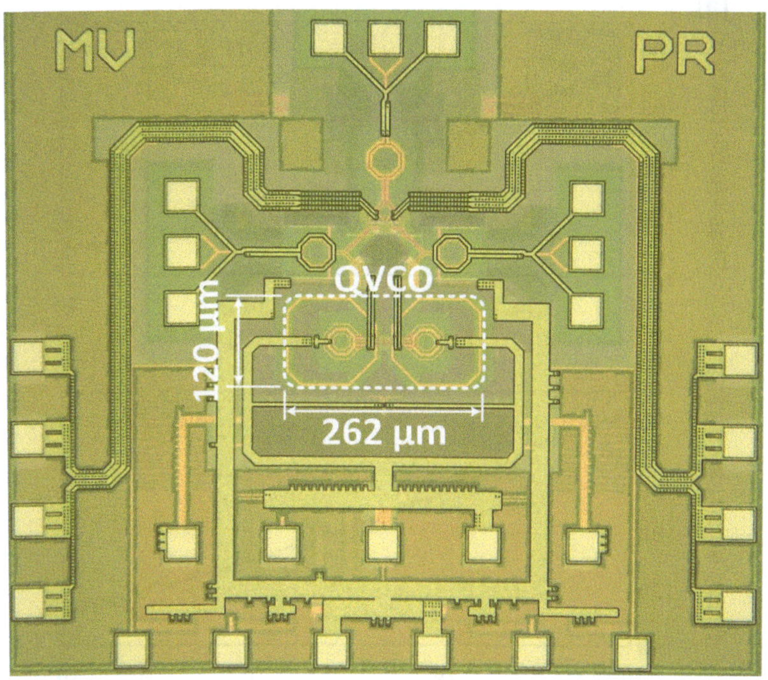

Fig. 4.30 Die micrograph of the realized test chip (core dimensions: 120 µm × 262 µm). ©2016 IEEE. Reprinted, with permission, from [49]

active area of only 0.031 mm^2. All the measurements are performed on an high frequency probe station. The mm-Wave output of the QVCO after the buffer and the external LO input of the double balanced I/Q mixer (see Fig. 4.29) are directly accessed by GSG probes, while the DC and IF signal pads are wire-bonded to a printed circuit board (PCB). The quadrature VCO consumes 35.6mW from a 0.7 V supply. The oscillation frequency is tunable from 71.4 to 76.1 GHz when M_{SW} is OFF and from 85.6 to 90.7 GHz when M_{SW} is ON, corresponding to 9.8 GHz of total tuning range. By varying the A-MOS varactor voltage from 0 to 1.2 V and acting on the two binary-weighted digitally controlled MOM capacitors, the oscillator realizes continuous tuning within the two bands.

Figure 4.31a, b show the measured phase noise from a 72.7 and 88.2 GHz carrier respectively. The signal is measured at the output of the buffer directly at mm-Wave and downconverted with an external mixer. The prototype achieves a measured phase noise at 1 and 10 MHz offset of −93.5 and −117.7 dBc/Hz from a 72.7 GHz carrier and −86.2 and −110 dBc/Hz from a 88.2 GHz carrier. The measured $1/f^3$ corner is ≈2 MHz. The same measurements are repeated over the tuning range are summarized and compared against simulations in Fig. 4.32a at 10 MHz offset, showing that the

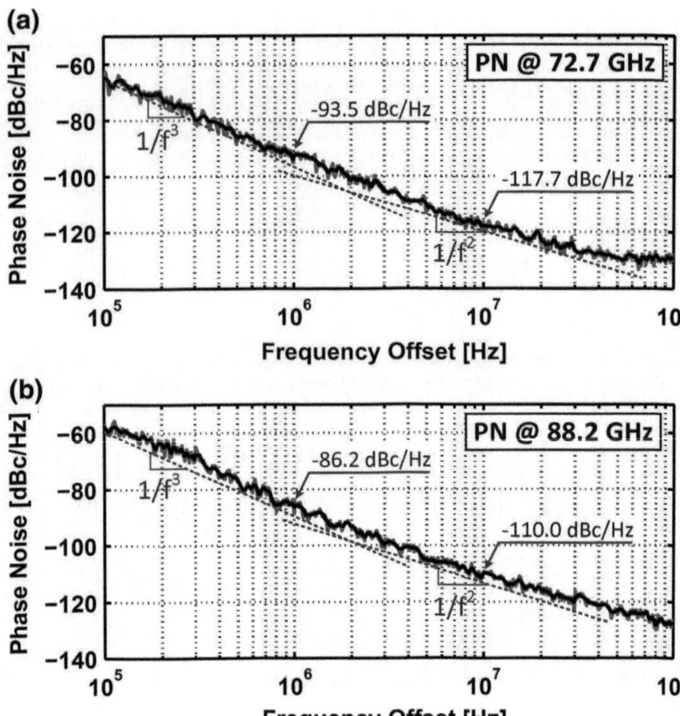

Fig. 4.31 Measured phase noise from a 72.7 GHz carrier (**a**) and from a 88.2 GHz carrier (**b**). ©2016 IEEE. Reprinted, with permission, from [49]

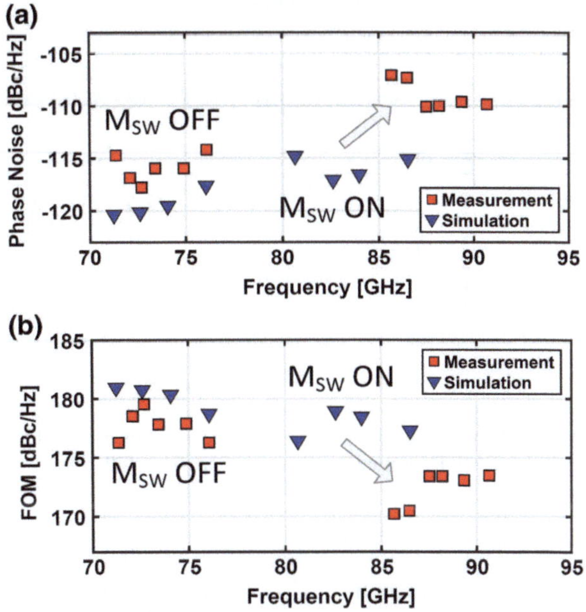

Fig. 4.32 Measured and simulated phase noise (**a**) and noise FOM (**b**) at 10 MHz offset from the carrier against frequency. ©2016 IEEE. Reprinted, with permission, from [49]

measured phase noise ranges from −114.2 to −117.7 dBc/Hz in the lower band and from −107 to −110 dBc/Hz in the higher one. The resulting measured noise figure of merit ranges from 176.3 to 179.4 dBc/Hz and from 170.2 to 173.4 dBc/Hz in the higher and lower band respectively, as reported in Fig. 4.32b together with the expected results. No RF transistor model was available during the design phase, resulting in an inaccurate estimation of the ON resistance of M_{SW}, much larger than the other transistors by design as explained in Sect. 4.3.2. The measured oscillation frequency in the higher band shows therefore a shift of about 4.6 GHz toward higher frequencies, giving rise to a deviation from the optimal design point and a degradation of phase noise performance in this band.

To measure the quadrature amplitude and phase imbalance, an external mm-Wave signal is applied to the on-chip I/Q mixer driven by the QVCO. The resulting down-converted IF outputs are then measured with a sampling oscilloscope and shown in Fig. 4.33. Measurements repeated over the whole tuning range prove a phase error less than 1.5° in the lower band and less than 3.5° in the higher one. The amplitude error stays always below 1 dB in both bands. Noteworthy, in a practical system relatively simple on-chip calibration techniques may be adopted to compensate for such a limited phase error, allowing high order modulation schemes as 64-QAM [63].

Table 4.1 summarizes and compares the measured performance of the quadrature VCO prototype to state-of-the-art integrated quadrature frequency generation circuits in the 70–100 GHz band. Benefited by the presented design techniques, this work achieves the lowest power consumption while occupying the smaller silicon area, and showing a better or comparable phase noise that varies less than 3.5 dB within each band.

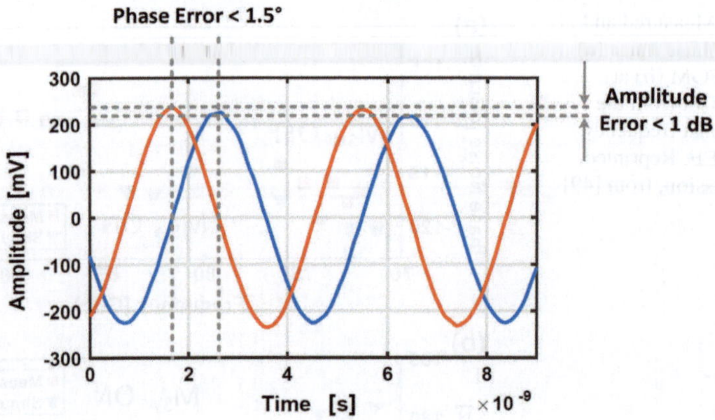

Fig. 4.33 Measured phase and amplitude imbalance of the I/Q signals downconverted to 260 MHz. ©2016 IEEE. Reprinted, with permission, from [49]

When such quadrature generation circuits are employed in direct conversion transceivers, the LO feedthrough and PA pulling may become serious issues [63–66]. It is therefore desirable to keep the number of on-chip inductors as small as possible and, in mm-Wave CMOS design, area serves as straightforward measure of this. Among the excellent designs in Table 4.1, this work stands out for the lowest reported silicon area, without trading in power consumption or phase noise performance, leading to a measured FOM_A between 3.6 and 12.8 dB higher than the best previously reported one.

4.3.4 Appendix

Assuming differential quadrature operation, the equivalent AC large signal model of the circuit schematic in Fig. 4.16 can be redrawn as in Fig. 4.34, where the ground reference is instrumentally shifted. Following the dissertation presented by Bevilacqua and Andreani in [9], in steady state is now possible to replace the transconductor with its describing function approximation [1]. The resulting circuit is shown in Fig. 4.35. By applying the Norton's theorem the current source $I_{\omega 0}$ can be substituted with an equivalent current source $\beta I_{\omega 0}$ in parallel with C_m as in Fig. 4.19, given

$$\beta = \frac{(sC_1 - jY_{12} + Y_{22})C_2 + C_1 Y_{22}}{(sC_1 + 2Y_{22})C_2 + C_1 Y_{22}}. \tag{4.46}$$

Since β is in general a complex number (i.e. the current flowing through the transistor $I_{\omega 0}$ and $\beta I_{\omega 0}$ are not in phase) and the transistor does enter the triode region as discussed in Sect. 4.3.1, the general result on phase noise stated in [3, 4]

Table 4.1 Comparison with state-of-the-art integrated quadrature frequency generation circuits in the 70/100 GHz band. ©2016 IEEE. Reprinted, with permission, from [49]

Ref.	**Topology**	**Freq. (GHz)**	**TR (GHz)**	**Power (mW)**	**PN@10 MHz (dBc/Hz)**	**FOM (dBc/Hz)**	**FOM_A (dBc/Hz)**	**Phase error**	**Area (mm^2)**	**Tech.**
[67]	VCO + RC PPF	70–89	19	310.2	−107/ −114(1)	159/168.1	168.7/177.8	<8.5°	0.107(2)	0.35µm SiGe
[37]	ILFM3	70.5–85.5	15	47.3	−111.7/ −115.8(1)	173.5/176.3	178.9/181.7	<2°	0.291(2)	65 nm CMOS
[68]	QVCO	90–94	4	43.2	−110.5(1)	173.5	n.a.	n.a.	n.a.	65 nm CMOS
This work [49]	**QVCO**	**71.4–76.1** **85.6–90.7**	**9.8**	**35.6**	**−114.2/−117.7** **−107/−110**	**176.3/179.4** **170.2/173.4**	**191.4/194.5** **185.3/188.5**	<1.5° <3.5°	**0.031**	**28 nm CMOS**

#Graphically estimated
+In-band best/worst

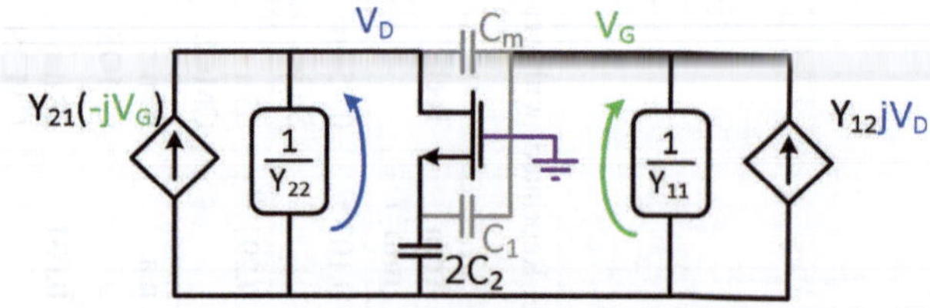

Fig. 4.34 Single ended AC large signal equivalent half circuit of the oscillator with an instrumental shift of the reference ground plane. ©2016 IEEE. Reprinted, with permission, from [49]

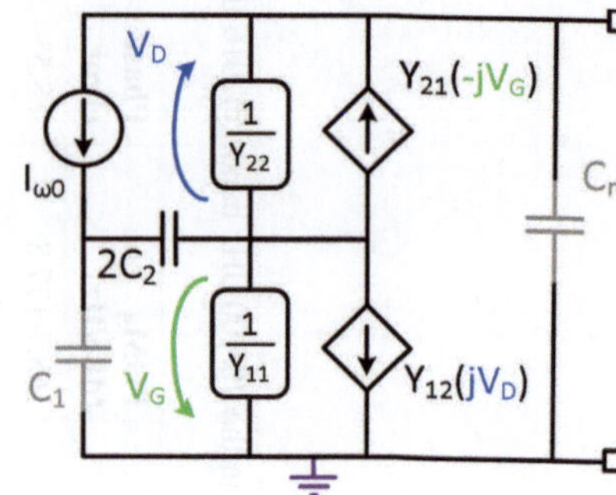

Fig. 4.35 Rearrangement of the circuit in Fig. 4.34 where the transconductor is replaced with its describing function approximation. ©2016 IEEE. Reprinted, with permission, from [49]

does not apply. Therefore, in this work this simplified model is only used to get insight into the quadrature operation of the circuit and obtain an approximated expression of the oscillation frequency.

4.4 Conclusion

This chapter discussed the fundamentals of low-noise *LC* CMOS oscillators. Simple equivalent circuit models, the general result on phase noise and circuit flicker noise to close-in phase noise upconversion mechanisms have been briefly recalled in Sect. 4.1, giving design insight into the operation of the circuit and highlighting the major challenges at mm-Wave. The most popular tuning extension techniques for *LC* VCOs and DCOs have been the focus of Sect. 4.2.

Section 4.3 presented a low-noise dual-band QVCO tailored for direct conversion E-Band transceivers. The prototype, implemented in 28 nm CMOS, adopts gate-to-drain transformers to couple two fundamental oscillators and generate accurate quadrature phases. Further, switched coupled inductors are added for tuning extension. Thanks to these techniques, it is possible to cover two bands with a single quadrature VCO, without jeopardizing phase noise or demanding extensive silicon area. The oscillator occupies a core area of only 0.031 mm^2 and is tunable from 71 to 76 GHz and 85.6 to 90.7 GHz, resulting in a total tuning range of 9.8 GHz. The peak phase noise at 10 MHz offset from the carrier is - 117.7 dBc/Hz in the lower band and −110 dBc/Hz in the higher one and varies less than 3.5 dB within each sub-band. The maximum phase error is 1.5 and 3.5 in the lower and higher band respectively.

Chapter 5 is dedicated to high frequency divider. Together with the QVCO, this block realizes the mm-Wave front-end of any analog or digital fundamental PLL and therefore limits the noise versus power performance of the whole system.

References

1. T.H. Lee, *The Design of CMOS Radio-Frequency Integrated Circuits* (Cambridge university press, Cambridge, 2003)
2. A. Hajimiri, T.H. Lee, A general theory of phase noise in electrical oscillators. IEEE J. Solid-State Circuits **33**(2), 179–194 (1998)
3. A. Mazzanti, P. Andreani, Class-C harmonic CMOS VCOs, with a general result on phase noise. IEEE J. Solid-State Circuits **43**(12), 2716–2729 (2008)
4. D. Murphy, J.J. Rael, A.A. Abidi, Phase noise in LC oscillators: a phasor-based analysis of a general result and of loaded Q. IEEE Trans. Circuits Syst. I Regul. Pap. **57**(6), 1187–1203 (2010)
5. Behzad Razavi, *RF Microelectronics*, 2nd edn. (Prentice Hall, New Jersey, 2011)
6. M. Garampazzi et al., An intuitive analysis of phase noise fundamental limits suitable for benchmarking LC oscillators. IEEE J. Solid-State Circuits **49**(3), 635–645 (2014)
7. L. Fanori, P. Andreani, Highly efficient class-C CMOS VCOs, including a comparison with class-B VCOs. IEEE J. Solid-State Circuits **48**(7), 1730–1740 (2013)
8. D.B. Leeson, A simple model of feedback oscillator noise spectrum. Proc. IEEE **54**(2), 329–330 (1966)
9. F. Padovan, M. Tiebout, K.L.R. Mertens, A. Bevilacqua, A. Neviani, Design of low-noise K-band SiGe bipolar VCOs: theory and implementation. IEEE Trans. Circuits Syst. I Regul. Pap. **62**(2), 607–615 (2015)
10. L. Romano, A. Bonfanti, S. Levantino, C. Samori, A.L. Lacaita, 5-GHz oscillator array with reduced flicker up-conversion in 0.13-μm CMOS. IEEE J. Solid-State Circuits **41**(11), 2457–2467 (2006)
11. S.A.R. Ahmadi-Mehr, M. Tohidian, R.B. Staszewski, Analysis and design of a multi-core oscillator for ultra-low phase noise. IEEE Trans. Circuits Syst. I Regul. Pap. **63**(4), 529–539 (2016)
12. W. Wu, R.B. Staszewski, J.R. Long, A 56.4-to-63.4 GHz multi-rate all-digital fractional-N PLL for FMCW radar applications in 65 nm CMOS. IEEE J. Solid-State Circuits **49**(5), 1081–1096 (2014)
13. S. Levantino, C. Samori, A. Zanchi, A.L. Lacaita, AM-to-PM conversion in varactor-tuned oscillators. IEEE Trans. Circuits Syst. II: Analog Digit. Signal Process. **49**(7), 509–513 (2002)
14. E. Hegazi, A.A. Abidi, Varactor characteristics, oscillator tuning curves, and AM-FM conversion. IEEE J. Solid-State Circuits **38**(6), 1033–1039 (2003)
15. A. Bevilacqua, P. Andreani, An analysis of 1/f noise to phase noise conversion in CMOS harmonic oscillators. IEEE Trans. Circuits Syst. I Regul. Pap. **59**(5), 938–945 (2012)
16. M. Shahmohammadi, M. Babaie, R.B. Staszewski, A 1/f noise upconversion reduction technique for voltage-biased RF CMOS oscillators. IEEE J. Solid-State Circuits **51**(11), 2610–2624 (2016)
17. F. Pepe, P. Andreani, A general theory of phase noise in transconductor-based harmonic oscillators. IEEE Trans. Circuits Syst. I Regul. Pap. **64**(2), 432–445 (2017)
18. E. Hegazi, H. Sjoland, A.A. Abidi, A filtering technique to lower LC oscillator phase noise. IEEE J. Solid-State Circuits **36**(12), 1921–1930 (2001)
19. D. Murphy, H. Darabi, H. Wu, 25.3 A VCO with implicit common-mode resonance, in *2015 IEEE International Solid-State Circuits Conference - (ISSCC) Digest of Technical Papers*, San Francisco, CA (2015), pp. 1–3

20. D. Murphy, H. Darabi, 2.5 A complementary VCO for IoE that achieves a 195dBc, Hz FOM and flicker noise corner of 200kHz, in *2016 IEEE International Solid-State Circuits Conference (ISSCC)*, San Francisco, CA (2016), pp. 44–45
21. M. Babaie, R.B. Staszewski, A class-F CMOS oscillator. IEEE J. Solid-State Circuits **48**(12), 3120–3133 (2013)
22. Huijung Kim, Seonghan Ryu, Yujin Chung, Jinsung Choi, Bumman Kim, A low phase-noise CMOS VCO with harmonic tuned LC tank. IEEE Trans. Microw. Theory Tech. **54**(7), 2917–2924 (2006)
23. C. Samori, Understanding phase noise in LC VCOs: a key problem in RF integrated circuits. IEEE Solid-State Circuits Mag. **8**(4), 81–91 (2016)
24. F. Pepe, P. Andreani, Still more on the $1/f^2$ phase noise performance of harmonic oscillators. IEEE Trans. Circuits Syst. II Express Briefs **63**(6), 538–542 (2016)
25. A. Moroni, R. Genesi, D. Manstretta, Analysis and design of a 54 GHz distributed hybrid wave oscillator array with quadrature outputs. IEEE J. Solid-State Circuits **49**(5), 1158–1172 (2014)
26. J. Wood, T.C. Edwards, S. Lipa, Rotary traveling-wave oscillator arrays: a new clock technology. IEEE J. Solid-State Circuits **36**(11), 1654–1665 (2001)
27. K. Takinami, R. Strandberg, P.C.P. Liang, G. Le Grand de Mercey, T. Wong, M. Hassibi, A distributed oscillator based all-digital PLL with a 32-phase embedded phase-to-digital converter. IEEE J. Solid-State Circuits **46**(11), 2650–2660 (2011)
28. A. Devos, M. Vigilante, P. Reynaert, Multiphase digitally controlled oscillator for future 5G phased arrays in 90 nm CMOS, in *2016 IEEE Nordic Circuits and Systems Conference (NORCAS)*, Copenhagen (2016), pp. 1–4
29. N. Nouri, J.F. Buckwalter, A 45-GHz rotary-wave voltage-controlled oscillator. IEEE Trans. Microw. Theory Tech. **59**(2), 383–392 (2011)
30. P. Kinget, Integrated GHz voltage controlled oscillators, *Analog Circuit Design* (Springer, US, 1999), pp. 353–381
31. B. Soltanian, H. Ainspan, W. Rhee, D. Friedman, P.R. Kinget, An ultra-compact differentially tuned 6-GHz CMOS LC-VCO with dynamic common-mode feedback. IEEE J. Solid-State Circuits **42**(8), 1635–1641 (2007)
32. L. Iotti, A. Mazzanti, F. Svelto, Insights into phase-noise scaling in switch-coupled multi-core LC VCOs for E-band adaptive modulation links. IEEE J. Solid-State Circuits **52**(7), 1703–1718 (2017)
33. B. Razavi, A 300-GHz fundamental oscillator in 65-nm CMOS technology. IEEE J. Solid-State Circuits **46**(4), 894–903 (2011)
34. C. Jany, A. Siligaris, J.L. Gonzalez-Jimenez, P. Vincent, P. Ferrari, A programmable frequency multiplier-by-29 architecture for millimeter wave applications. IEEE J. Solid-State Circuits **50**(7), 1669–1679 (2015)
35. P. Reynaert, W. Steyaert, M. Vigilante, "RF CMOS". Nanoelectronics: Materials, Devices, Applications, 2 Volumes (2017)
36. E. Mammei, E. Monaco, A. Mazzanti, F. Svelto, A 33.6-to-46.2GHz 32nm CMOS VCO with 177.5dBc, Hz minimum noise FOM using inductor splitting for tuning extension, in *2013 IEEE International Solid-State Circuits Conference Digest of Technical Papers*, San Francisco, CA (2013), pp. 350–351
37. Z. Huang, H.C. Luong, B. Chi, Z. Wang, H. Jia, 25.6 A 70.5-to-85.5GHz 65nm phase-locked loop with passive scaling of loop filter, in *2015 IEEE International Solid-State Circuits Conference - (ISSCC) Digest of Technical Papers*, San Francisco, CA (2015), pp. 1–3
38. H. Jia, L. Kuang, Z. Wang, B. Chi, A W-band injection-locked frequency doubler based on top-injected coupled resonator. IEEE Trans. Microw. Theory Tech. **64**(1), 210–218 (2016)
39. A.H. Masnadi Shirazi et al., On the design of mm-wave self-mixing-VCO architecture for high tuning-range and low phase noise. IEEE J. Solid-State Circuits **51**(5), 1210–1222 (2016)
40. Z. Zong, M. Babaie, R.B. Staszewski, A 60 GHz frequency generator based on a 20 GHz oscillator and an implicit multiplier. IEEE J. Solid-State Circuits **51**(5), 1261–1273 (2016)
41. M. Demirkan, S.P. Bruss, R.R. Spencer, Design of wide tuning-range CMOS VCOs using switched coupled-inductors. IEEE J. Solid-State Circuits **43**(5), 1156–1163 (2008)

42. T. LaRocca, J.Y.C. Liu, M.C.F. Chang, 60 GHz CMOS amplifiers using transformer-coupling and artificial dielectric differential transmission lines for compact design. IEEE J. Solid-State Circuits **44**(5), 1425–1435 (2009)
43. T. LaRocca, J. Liu, F. Wang, F. Chang, Embedded DiCAD linear phase shifter for 5765GHz reconfigurable direct frequency modulation in 90nm CMOS, in *2009 IEEE Radio Frequency Integrated Circuits Symposium*, Boston, MA (2009), pp. 219–222
44. T. LaRocca, J. Liu, F. Wang, D. Murphy, F. Chang, CMOS digital controlled oscillator with embedded DiCAD resonator for 5864GHz linear frequency tuning and low phase noise, in *2009 IEEE MTT-S International Microwave Symposium Digest*, Boston, MA (2009), pp. 685–688
45. W. Wu, J.R. Long, R.B. Staszewski, High-resolution millimeter-wave digitally controlled oscillators with reconfigurable passive resonators. IEEE J. Solid-State Circuits **48**(11), 2785–2794 (2013)
46. A. Bevilacqua, F.P. Pavan, C. Sandner, A. Gerosa, A. Neviani, Transformer-based dual-mode voltage-controlled oscillators. IEEE Trans. Circuits Syst. II Express Briefs **54**(4), 293–297 (2007)
47. J. Yin, H.C. Luong, A 57.590.1-GHz magnetically tuned multimode CMOS VCO. IEEE J. Solid-State Circuits **48**(8), 1851–1861 (2013)
48. A. Mazzanti, A. Bevilacqua, On the phase noise performance of transformer-based CMOS differential-pair harmonic oscillators. IEEE Trans. Circuits Syst. I Regul. Pap. **62**(9), 2334–2341 (2015)
49. M. Vigilante, P. Reynaert, Analysis and design of an E-band transformer-coupled low-noise quadrature VCO in 28-nm CMOS. IEEE Trans. Microw. Theory Tech. **64**(4), 1122–1132 (2016)
50. L. Li, P. Reynaert, M. Steyaert, A colpitts LC VCO with Miller-capacitance gm enhancing and phase noise reduction techniques, in *2011 Proceedings of the ESSCIRC (ESSCIRC)*, Helsinki (2011), pp. 491–494
51. M.M. Bajestan, V.D. Rezaei, K. Entesari, A low phase-noise wide tuning-range quadrature oscillator using a transformer-based dual-resonance LC ring. IEEE Trans. Microw. Theory Tech. **63**(4), 1142–1153 (2015)
52. A. Bevilacqua, F.P. Pavan, C. Sandner, A. Gerosa, A. Neviani, A 3.4-7 GHz transformer-based dual-mode wideband VCO, in *2006 Proceedings of the 32nd European Solid-State Circuits Conference*, Montreux (2006), pp. 440–443
53. G. Li, L. Liu, Y. Tang, E. Afshari, A low-phase-noise wide-tuning-range oscillator based on resonant mode switching. IEEE J. Solid-State Circuits **47**(6), 1295–1308 (2012)
54. S. Levantino, P. Maffezzoni, F. Pepe, A. Bonfanti, C. Samori, A.L. Lacaita, Efficient calculation of the impulse sensitivity function in oscillators. IEEE Trans. Circuits Syst. II Express Briefs **59**(10), 628–632 (2012)
55. M. Babaie, R.B. Staszewski, An ultra-low phase noise class-F 2 CMOS oscillator with 191 dBc/Hz FoM and long-term reliability. IEEE J. Solid-State Circuits **50**(3), 679–692 (2015)
56. D. Murphy et al., A low phase noise, wideband and compact CMOS PLL for use in a heterodyne 802.15.3c transceiver. IEEE J. Solid-State Circuits **46**(7), 1606–1617 (2011)
57. A. Mazzanti, F. Svelto, P. Andreani, On the amplitude and phase errors of quadrature LC-tank CMOS oscillators. IEEE J. Solid-State Circuits **41**(6), 1305–1313 (2006)
58. N.C. Kuo, J.C. Chien, A.M. Niknejad, Design and analysis on bidirectionally and passively coupled QVCO with nonlinear coupler. IEEE Trans. Microw. Theory Tech. **63**(4), 1130–1141 (2015)
59. W. Sansen, 1.3 Analog CMOS from 5 micrometer to 5 nanometer, in *2015 IEEE International Solid-State Circuits Conference - (ISSCC) Digest of Technical Papers*, San Francisco, CA (2015), pp. 1–6
60. D. Zhao, P. Reynaert, A 60-GHz dual-mode class AB power amplifier in 40-nm CMOS. IEEE J. Solid-State Circuits **48**(10), 2323–2337 (2013)
61. J. Shi, K. Kang, Y.Z. Xiong, J. Brinkhoff, F. Lin, X.J. Yuan, Millimeter-wave passives in 45-nm digital CMOS. IEEE Electron Device Lett. **31**(10), 1080–1082 (2010)
62. U. Decanis, A. Ghilioni, E. Monaco, A. Mazzanti, F. Svelto, A low-noise quadrature VCO based on magnetically coupled resonators and a wideband frequency divider at millimeter waves. IEEE J. Solid-State Circuits **46**(12), 2943–2955 (2011)

63. D. Zhao, P. Reynaert, A 40 nm CMOS E-band transmitter with compact and symmetrical layout floor plans. IEEE J. Solid-State Circuits **50**(11), 2560–2571 (2015)
64. A. Mirzaei, M. Mikhemar, H. Darabi, 21.8 A pulling mitigation technique for direct-conversion transmitters, in *2014 IEEE International Solid-State Circuits Conference Digest of Technical Papers (ISSCC)*, San Francisco, CA (2014), pp. 374–375
65. B. Razavi, A study of injection locking and pulling in oscillators. IEEE J. Solid-State Circuits **39**(9), 1415–1424 (2004)
66. B. Razavi, Design considerations for direct-conversion receivers. IEEE Trans. Circuits Syst. II: Analog Digit. Signal Process. **44**(6), 428–435 (1997)
67. I. Nasr, B. Laemmle, K. Aufinger, G. Fischer, R. Weigel, D. Kissinger, A 70–90-GHz high-linearity multi-band quadrature receiver in 0.35μ m SiGe technology. IEEE Trans. Microw. Theory Tech. **61**(12), 4600–4612 (2013)
68. E. Laskin et al., Nanoscale CMOS transceiver design in the 90170-GHz range. IEEE Trans. Microw. Theory Tech. **57**(12), 3477–3490 (2009)

Chapter 5
mm-Wave Dividers

The PLL is a key subsystems of any transceiver for wireless applications. In state-of-the-art fundamental mm-Wave PLL (both analog and digital) the first divider and oscillator run at the highest frequency, becoming the system bottleneck for noise, tuning-range, power consumption and yield under PVT variation [1, 2]. It is therefore highly desirable to adopt robust low power solutions for the frequency divider, with possibly a large tuning capability to overcome the variation.

Injection locked (IL) *LC* frequency dividers achieve the higher speed for a given power consumption but need one or even more on-chip inductors rising the complexity of the design and yielding a large area consumption for a limited locking range (LR) [3–7]. Static CML dividers, on the other hand, are famous for the wide LR, but require a large power consumption to work at high speed, even if inductive peaking techniques are used [8]. In [9] an *RC* static divider based on CML dynamic latches with load modulation is proposed. This topology, derived by the traditional CML static one, improves the divider performance at high frequencies, leading to a low power tunable solution.

This chapter is organized as follow. The basic concept of injection locking is revised in Sect. 5.1. This technique is particularly powerful and commonly adopted by many state-of-the-art high speed low power frequency dividers and multipliers. It is also useful to study the effect of coupled oscillators (such as quadrature VCOs) and the undesired effect of pulling between two VCOs running at different frequencies on the same chip and/or between the VCO and the power amplifier in a direct conversion transmitter [10–12]. Section 5.2 recalls the most popular circuits used in state-of-the-art high speed dividers. The operation principle of each solution is briefly summarized and the design trade-offs are highlighted. Section 5.3 presents a systematic design methodology to maximize performance of *RC* static divider based on CML dynamic latches with load modulation in the frequency band from 60 to 90 GHz. A divide-by-4 prototype 28 nm bulk CMOS based on the proposed design techniques is fully

M. Vigilante and P. Reynaert, *5G and E-Band Communication Circuits in Deep-Scaled CMOS*, Analog Circuits and Signal Processing,
https://doi.org/10.1007/978-3-319-72646-5_5

characterized, demonstrating a measured operating range from 25 to 102 GHz, when drawing 2.81–3.04 mW from a 0.9 V supply.

The results of the mm-Wave frequency divider in Sect. 5.3 have been published in IEEE Asian Solid-State Circuits Conference (A-SSCC 2015).

5.1 Injection Locking: Operation Principle

Many systems in nature show a cyclostationary behavior (e.g. day and night Earth cycle, animals awaking/sleeping time, mechanical/electrical oscillators, etc.). It has been observed that such systems are prone to injection locking [10]. In the context of this work, such phenomenon is particularly relevant.

The basic idea is to realize a "lousy" free running oscillator (i.e. low power but highly noisy) and lock it with a much purer tone. When the system is properly locked, the resulting phase noise is imposed by the injected signal and we are left with a low power and low noise circuit. A number of questions rise. (1) How strong should the injected signal be to realize locking? (2) How far from its self-resonant frequency can an oscillator be locked? (3) How can we use such principle to realize frequency dividers/multipliers?

To get deeper insight and answer these questions, it is useful to refer to the negative G_m model of an injection locked LC oscillator in Fig. 5.1a. Where

$$V_T = Z_T I_T = Z_T (I_{Gm} + I_{inj}). \tag{5.1}$$

Two key observations are the following. (1) I_{Gm} is always in-phase with V_T since G_m is a real number. (2) As clear in Fig. 5.1b, when $f_{inj} \neq f_o$,

$$\angle Z_T = \angle \left(\frac{V_T}{I_T} \right) = -\varphi = \angle I_{Gm} - \angle I_T. \tag{5.2}$$

To realize locking therefore, the injected current I_{inj} should provide the required phase shift as depicted in the phasor diagram in Fig. 5.1c. For a given magnitude $|I_{inj}| < |I_{Gm}|$, the maximum phase shift that can be compensated is

$$\varphi_{MAX} = arcsin \left(\frac{|I_{inj}|}{|I_{Gm}|} \right) = arcsin \left(\frac{|I_{inj}|}{G_m |V_T|} \right). \tag{5.3}$$

If for a given frequency f_{inj} the required phase shift $|\angle Z_T| > \varphi_{MAX}$, the locking condition in Eq. 5.3 is not satisfied and the circuit can not be locked.

When an RLC tank is considered as in Fig. 5.1a, starting from the condition in Eqs. 5.2 and 5.3 it is possible to express the locking range (LR) as function of design parameters [10]

$$LR = \frac{2\pi f_o}{Q} \frac{I_{inj}}{I_{Gm}} \frac{1}{\sqrt{1 - (I_{inj}/I_{Gm})^2}}. \tag{5.4}$$

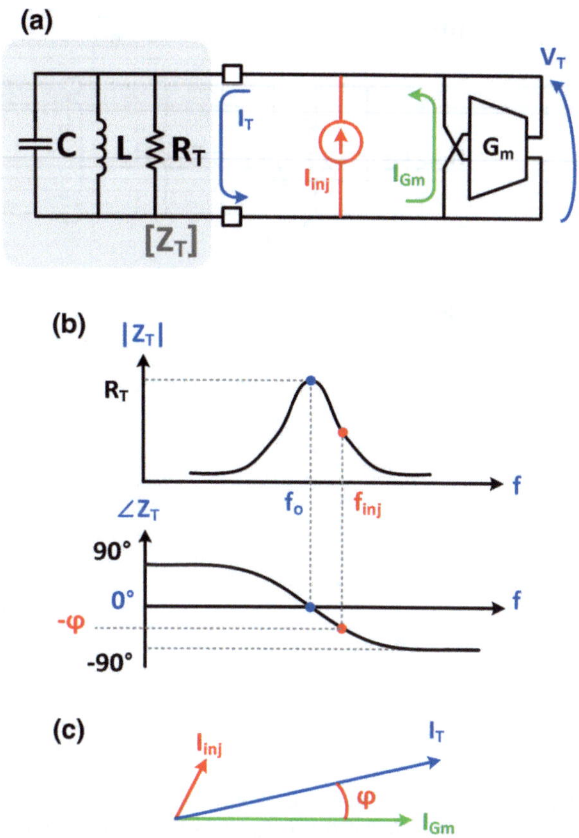

Fig. 5.1 Negative G_m model of a *LC* oscillator under injection locking (**a**), magnitude and phase of the tank impedance Z_T against frequency (**b**), and phasor diagram of the currents (**c**)

As clear from Eq. 5.4, to extend the LR of an injection locked oscillator loaded with an *RLC* tank a designer can work on three main parameters. (1) Increasing the injected current I_{inj}. Normally at expense of a higher power consumption. (2) Reducing $I_{Gm} = G_m V_T$, by lowering G_m. However, a minimum G_m is still required to compensate for the tank losses and guarantee the start-up conditions under PVT variation. (3) Lowering the Q-factor of the tank at the expenses of higher DC power consumption.

To further extend the LR a 4th order tank can be adopted [14, 15]. Figure 5.2 shows the schematic and the input impedance Z_{11} of such network and compares it against the classical *RLC* tank. Here $C_1 = C_2, L_1 = L_2, R_1 = R_2, k = 0.2$ and $Q = 5$ are assumed. Clearly, the condition $|\angle Z_{11}| < \varphi_{MAX}$ is met over a larger frequency range when a well designed 4th order tank is adopted (i.e. effectively enlarging the LR). Elaborating further on this concept, a 6th order tank based on a three-winding transformer has been proposed in [16].

In is noteworthy to remark that this result is general for any injection locked oscillator. Therefore even if the techniques described in [14–16] where developed

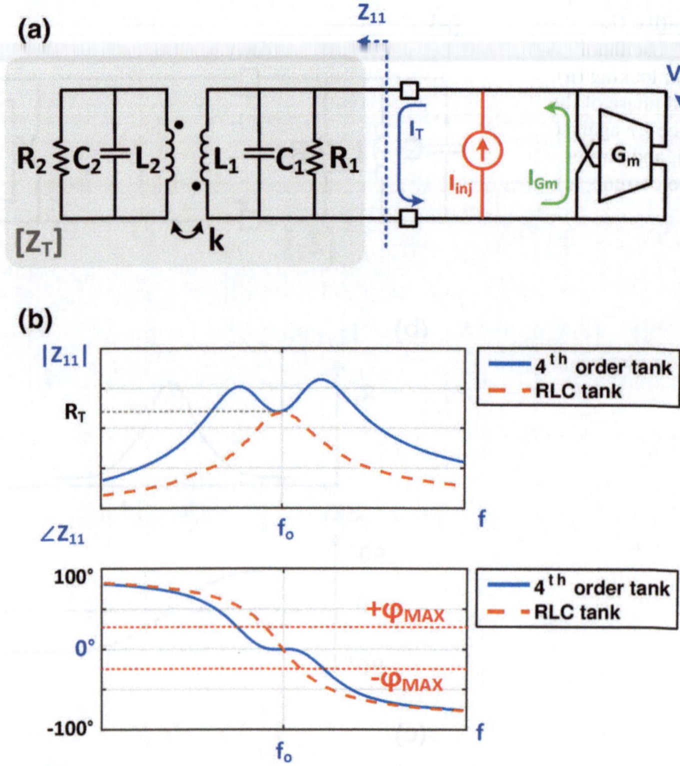

Fig. 5.2 Negative G_m model of an injection locked oscillator with a 4th order tank (**a**), magnitude and phase of the tank impedance Z_{11} against frequency and comparison with the classical *RLC* tank (**b**) [14]

for IL frequency multipliers, the very same applies for IL frequency dividers. The difference stands in the injection mechanism. In a frequency multiplier, the n-th harmonic of a low noise low frequency reference is injected. Whereas, in a divider the non-linearity of the injection devices are used to realize a mixer and inject a 1/n downconverted input signal.

5.2 High Speed Dividers

5.2.1 Injection Locked LC Dividers

Figure 5.3 shows the conceptual schematic of an injection locked frequency divide-by-4 (ILFD4). An harmonic mixer is used to up-convert the tank voltage to $3f_{inj}$ and multiply it with the input signal at $4f_{inj}$. The current at the output of the harmonic

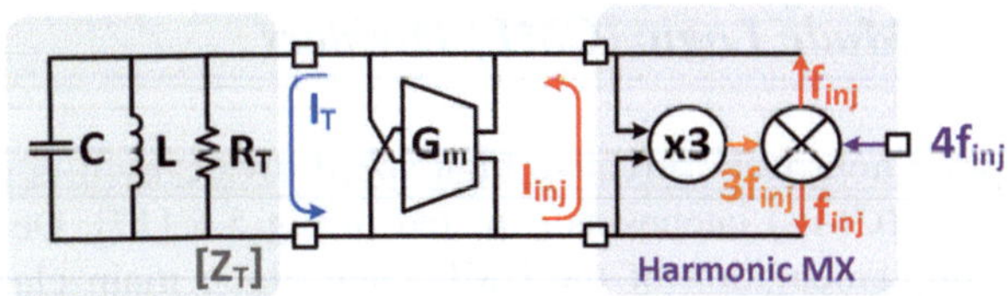

Fig. 5.3 Conceptual schematic of an injection locked frequency divide-by-4

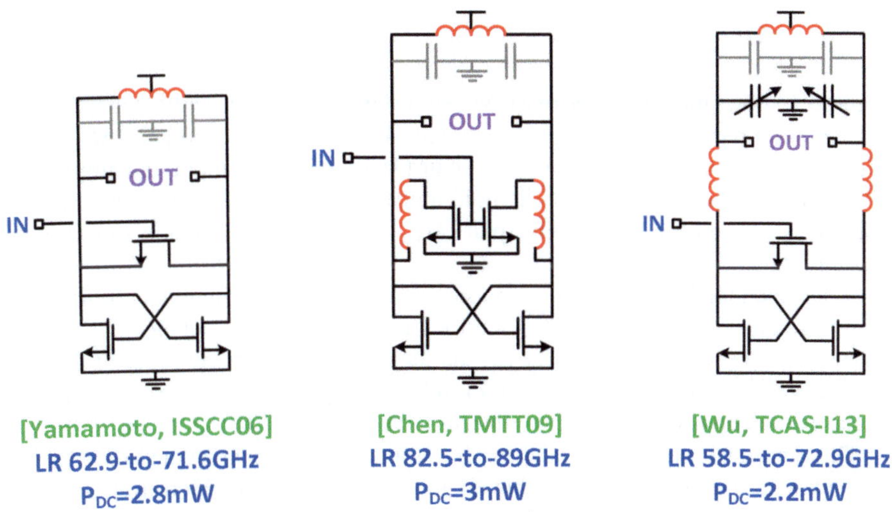

Fig. 5.4 State-of-the-art injection locked frequency divide-by-4 implementation, schematic and main performance parameters presented in **a** [3], **b** [4] and **c** [6]

mixer is then filtered by the tank and only the desired component at f_{inj} survives, realizing the desired frequency division ratio.

A simple and effective implementation of such divider is shown in Fig. 5.4a. Here the non-linearity of a single NMOS devices is used to realize the harmonic mixer [3]. To further improve the performance of such ILFD4, the circuits in Fig. 5.4b, c have been proposed in [4, 6].

These dividers achieve high frequency of operation at extremely low power consumption. However, they suffer from limited tuning capabilities and need 1–3 on-chip inductors. The design of such dividers is therefore particularly involved, especially when implemented in a real system. Where the divider needs to be able to track the oscillator over the required tuning range under PVT variations. Moreover, the need for inductors calls for high area consumption and makes the divider sensitive to spurious magnetic coupling from and to other circuits. Finally, Since injection locked frequency dividers rely on IL oscillators, they share similar trade-offs between power consumption and tuning range and suffer the same limitations coming from technology scaling and high frequency operation discussed in Chaps. 2 and 4.

5.2.2 *Current-Mode Logic (CML) Dividers*

A different popular approach to realize high speed divide-by-4 is to cascade four current-mode logic (CML) latches as depicted in Fig. 5.5 [17]. Depending on the specific circuit implementation of the CML latch, three main classes have been proposed in literature and are discussed in the following.

5.2.2.1 CML Static Dividers

Figure 5.6 shows the schematic of a CML static latch. It comprise a *RC* load (R_L, C_L), a differential pair (M_{SW}), a regenerative pair (M_{reg}) and a clocked pair (M_{B1}, M_{B2}). When the clock is high, the differential pair (M_{SW}) sense the input and amplify the difference between D and $\overline{D}$. In the hold mode, the clock goes down, the differential pair is off and the regenerative pair (M_{reg}) keeps the information stored in the capacitors C_L.

When such latch is used in the loop shown in Fig. 5.5, the circuit can be seen as a four-stage clocked ring oscillator. Therefore, even without an input clock, the circuit shows a self-oscillation frequency f_{SO} that depends on the *RC* product of the load and the bias current. If the voltage amplitude of the input clock is large enough, the circuit is injection-locked and functions as a divide-by-4 [17].

This circuit solution is best in terms of tuning capabilities when the load is implemented with a PMOS transistor biased in the triode region. Moreover, since no on-chip inductors are needed, the resulting silicon area is particularly compact. In

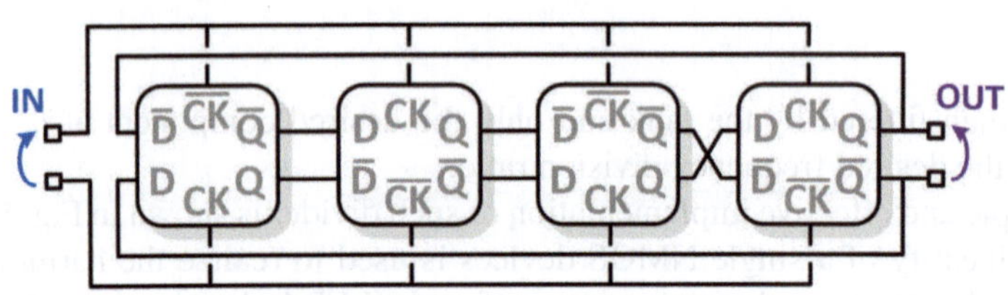

Fig. 5.5 Divide-by-4 based on CML latches block diagram

Fig. 5.6 CML static latch schematic

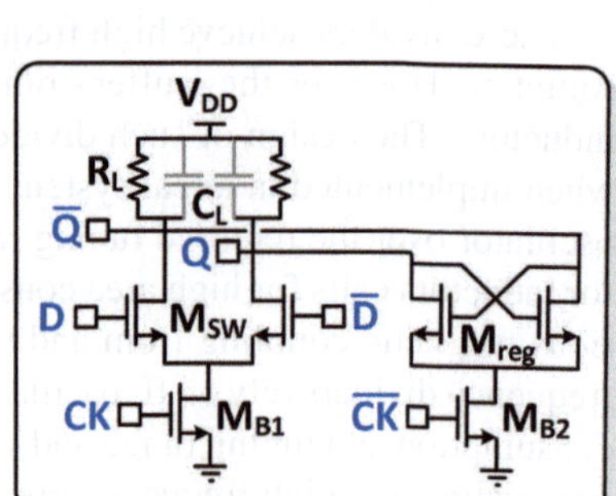

[18] a 94 GHz divide-by-2 in 65 nm SOI CMOS has been presented, proving that the speed of such circuit fully benefits from technology scaling.

5.2.2.2 CML Static Dividers with Inductive Peaking

To further increase the speed of this type of dividers, inductive peaking may be used as shown in Fig. 5.7 [8, 17]. When no clock is applied, the circuit clearly reduces to a *LC* oscillator. The loop in Fig. 5.5 therefore shows a behavior somewhere in between the CML static divider and an IL divide-by-4. When a larger inductor is used, the self oscillation frequency rises, but the circuit fails to lock at lower frequencies (Fig. 5.7).

This solution operates at higher frequency with a lower power consumption, but the advantage in terms of silicon area and tuning capabilities is lost.

5.2.2.3 Dynamic CML Latch Based Dividers

Figure 5.8 shows the schematic of the CML dynamic latch with load modulation proposed by Ghilioni et al. in [9]. When four of such latches are closed in the feedback loop of Fig. 5.5, the circuit shows an improved maximum operation frequency (f_{max}) and locking range when compared to the conventional divider based on static CML latches for a given power consumption. The key intuition is that, when the frequency of operation is high enough, the information stored in the loading capacitor C_L does not have time to leak through R_L and the regenerative pair of Fig. 5.6 might be removed, further reducing the parasitic capacitive loading.

The latch depicted in Fig. 5.8 consists of a differential pair where a complementary clock drives both the tail current source (M_1) and the PMOS load (M_3) biased in the triode region. The operation of the circuit can be divided in two regions: (1) sense

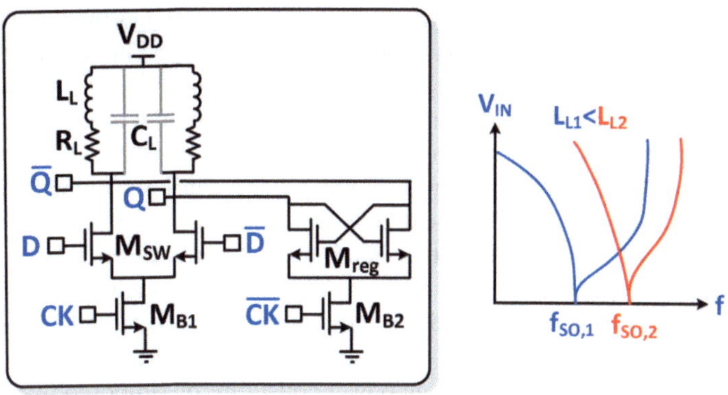

Fig. 5.7 CML static latch with inductive peaking schematic (left) and input sensitivity curves dependency on L_L (right) [17]

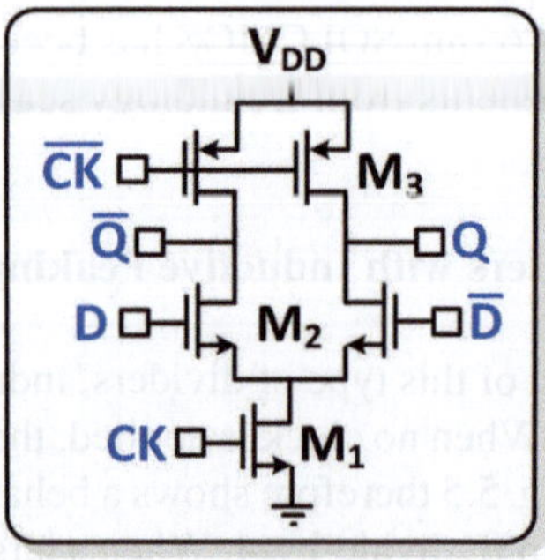

Fig. 5.8 CML dynamic latch with load modulation schematic

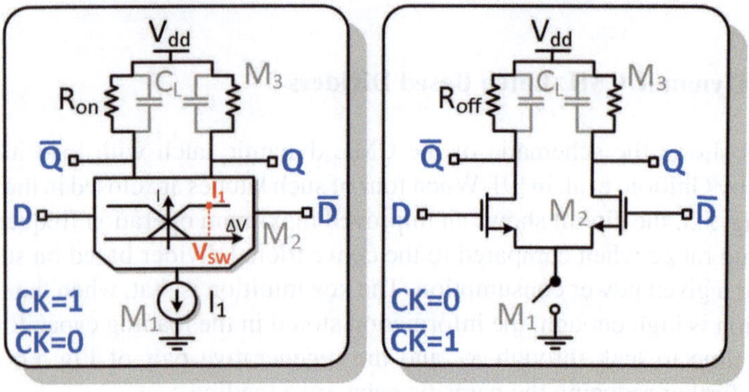

Fig. 5.9 AC large signal model of the CML dynamic latch with load modulation during the sense mode (left) and during the hold mode (right). ©2015 IEEE. Reprinted, with permission, from [21]

mode when CK = 1 and (2) hold mode when CK = 0. The AC large signal model of the circuit during the two different states is shown in Fig. 5.9. During the sense mode, the differential pair charges the load parasitic capacitor C_L. The differential output voltage tends asymptotically to $R_{on}I_1$ with a time constant $R_{on}C_L$. A low value of $R_{on}C_L$ together with a large bias current is therefore beneficial to increase the speed of the latch during this phase. When the output differential voltage reaches V_{SW} (see Fig. 5.9), the differential pair of the following stage can be switched completely, ensuring the correct operation of the divider. This threshold sets the limit for f_{max}.

During the hold phase, on the other hand, the tail current source is switched off and the parasitic capacitance at the output is discharged through R_{off}. To ensure the correct operation in this phase, the differential output voltage should not drop below V_{SW}. Hence, a large value of $R_{off}C_L$ is desirable to extend the hold phase, setting a lower bound for f_{min}. Clearly, for given power consumption (imposed by I_1), W_3 sets the ratio R_{off}/R_{on} yielding a trade-off between LR and maximum operation frequency. By the same token, increasing W_2 lowers V_{SW} at the cost of higher parasitic loading capacitance.

Under the simplifying assumption of the square-law operation region for the transconductors ($V_{SW} = \sqrt{2}(V_{GS} - V_T)$) and $R_{off} \rightarrow +\infty$, it is possible to express f_{max} as [9]

$$f_{max} = \frac{1}{2R_{on}C_L(1.41 + 0.59V_{SW}/(R_{on}I_1))\sqrt{V_{SW}/(R_{on}I_1)}}. \tag{5.5}$$

Clearly, to higher the frequency of operation the RC product of the load $R_{on}C_L$ shold be minimized. Or, for a given $R_{on}C_L$, $V_{SW}/(R_{on}I_1)$ should be minimized.

When the assumption $R_{off} \rightarrow +\infty$ is removed, the minimum frequency of operation can be found as [9]

$$f_{min} = \frac{1}{2R_{off}C_L\, ln(R_{on}I_1/V_{SW})}. \tag{5.6}$$

As expected, a large value of R_{off} is beneficial to f_{min}, while the product $V_{SW}/(R_{on}I_1)$ has a mild effect.

The ratio f_{max}/f_{min} can now be expressed as [9]

$$\frac{f_{max}}{f_{min}} = \frac{R_{off}}{R_{on}} \frac{ln(R_{on}I_1/V_{SW})}{(1.41 + 0.59V_{SW}/(R_{on}I_1))\sqrt{V_{SW}/(R_{on}I_1)}}. \tag{5.7}$$

From Eq. 5.7 is evident that a large ratio R_{off}/R_{on} is needed to enlarge the locking range, whereas C_L has no influence. Moreover, the larger the ratio $V_{SW}/(R_{on}I_1)$, the better.

Although this analysis is fundamental to get insight into the operation of the divider, it is important to remember that the Shichman-Hodges model holds true only for a narrow operation region in deep-scaled CMOS and below 20 nm this model has completely disappeared [13]. Moreover, not much has been reported so far about how the size and the bias point of M_1 affect the overall performance.

5.3 Design Example: An Ultra-wideband Divide-by-4 in 28nm CMOS

In this section a wideband tunable divide-by-4 is designed and realized in 28 nm bulk CMOS. A systematic design methodology to maximize the locking range over power consumption ratio is proposed. The test chip core area is only $25.6 \times 24.8\,\mu m^2$ and measurements repeated over several samples demonstrate an operating frequency range from 25 to 102 GHz with a maximum power consumption of 5.64 mW from a 0.9 V supply. The frequency band from 44.3 to 90 GHz is covered in only three steps with a minimum fractional bandwidth in exceed of 20% and power consumption less than 4.7 mW demonstrating the effectiveness of the proposed design techniques.

5.3.1 Design for Maximum Locking Range and Minimum Power Consumption in the E-Band

Figure 5.10 shows the schematic of the designed divide-by-4 highlighting the AC coupling, DC bias and transistor widths. To achieve optimal performance in the frequency band from 60 to 90 GHz, in this section a systematic design procedure is proposed. From the qualitative analysis given in Sect. 5.2, it is clear that a high frequency of operation comes at the cost of a narrow locking range for the given power consumption. The optimal design is therefore the one that maximize the following Figure of Merit:

$$FOM = \frac{LR}{P_{DC}}. \tag{5.8}$$

If no RF transistor model is available during the design phase, to avoid misleading results it is mandatory to include from an early stage of the design the layout parasitics due to the low metal connections to the transistor. After a first investigation of the divider operation adopting digital transistor models at a schematic level, a first estimation of the device dimension is obtained. Then, different transistor layouts are drawn and a post layout parasitic extractor tool (e.g. PEX, QRC, etc.) is used to predict their effect on the device performance. All active devices are designed with minimal length for maximum speed (L = 28 nm). To characterize the quantitative behavior of the divider, the expected FOM is investigated for different device sizes. For each con-

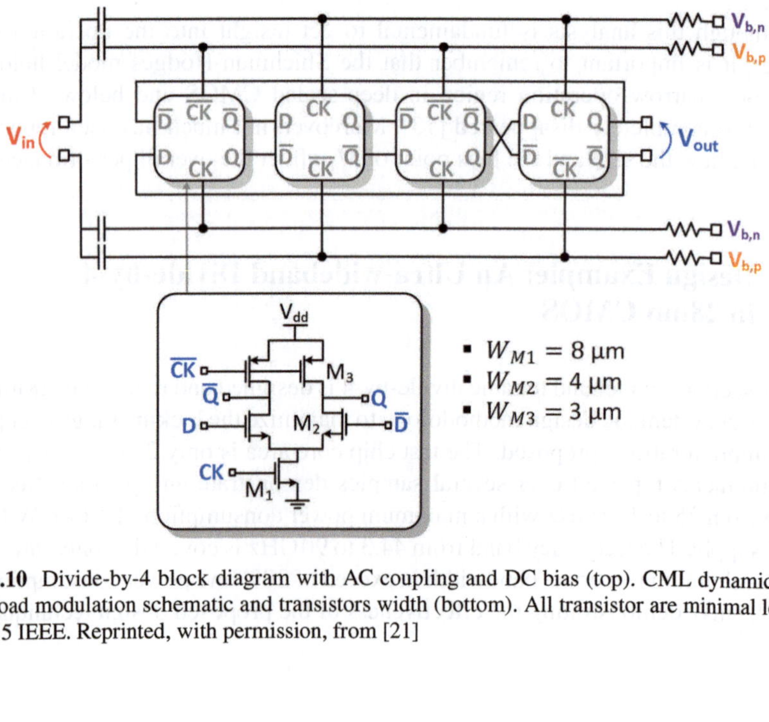

Fig. 5.10 Divide-by-4 block diagram with AC coupling and DC bias (top). CML dynamic latch with load modulation schematic and transistors width (bottom). All transistor are minimal length. ©2015 IEEE. Reprinted, with permission, from [21]

figuration the self oscillation frequency (f_{SO}) is set at about 80 GHz by acting on bias current I_1 and the bias point of the loading transistor, through $V_{b,n}$ and $V_{b,p}$ respectively (see Fig. 5.10). In Fig. 5.11 the expected FOM is reported against the PMOS loading transistor width (W_3) for different sizes of the differential pair (W_2), when the divider is driven by a differential sinusoidal clock with amplitude 400 mV 0-pk. The experiment is repeated for different sizes of the tail transistor, namely $W_1 = 6\,\mu m$ (Fig. 5.11a), $W_1 = 8\,\mu m$ (Fig. 5.11b), $W_1 = 10\,\mu m$ (Fig. 5.11c). The LR/P_{DC} ratio improves when M_1 and M_2 are designed with a relatively small width. Moreover, in Sect. 5.2 an optimum value of R_{off}/R_{on} was expected. This sweet spot is evident in Fig. 5.11a, b for a value of $W_3 = 3\,\mu m$. The downside of reducing the width of M_1 and M_2 is that, for the same frequency of operation, the output voltage swing is reduced. When $W_2 = 4\,\mu m$ and $W_3 = 3\,\mu m$ the simulated differential voltage swing 0-pk is 715, 655 and 530 mV for $W_1 = 10\,\mu m$, $W_1 = 8\,\mu m$ and $W_1 = 6\,\mu m$ respectively. Further decreasing W_1 to 4 µm leads to a 200 mV 0-pk voltage swing and a drop of f_{max} in favor of a reduction in power consumption. This case is therefore not reported in Fig. 5.11. To account for possible device model inaccuracy, in this design the transistor are overdesigned to $W_1 = 8\,\mu m$, $W_2 = 4\,\mu m$ and $W_3 = 3\,\mu m$ to ensure wide margin of operation.

5.3.2 Measurement Results

The die photograph of the divide-by-4 prototype realized in 28 nm bulk CMOS technology is shown in Fig. 5.12. Since this topology does not rely on on-chip inductors, the resulting core area is only $25.6 \times 24.8\,\mu m^2$. The test chip and measurement setup block diagram is shown in Fig. 5.13. Measurements are performed on a high-frequency probe station. The DC pads are wire-bonded to a PCB while the input and output pads are accessed by 50Ω GSG probes. To demonstrate the wideband operation of the designed prototype at mm-Wave with the available measurement equipment, the spectrum is divided in three parts. An E8257D Agilent PSG is used to generate the input signal up to 67 GHz. Two different source modules followed by a dedicated linear level set attenuator cover the band from 60 to 90 GHz and from 90 to 140 GHz respectively. For testing purpose, the input clock signal is applied to an on-chip transformer that serves as balun. A buffer is also implemented on-chip to drive the 50Ω measurement equipment. The output spectrum and phase noise is measured directly with a 43.5 GHz Rohde and Schwarz FSW Signal and Spectrum Analyzer.

Several samples are fully characterized demonstrating a broad frequency of operation and negligible difference among each other. All the experiments are performed with a 0.9 V supply. Figure 5.14 shows the measured self oscillation frequency versus DC power consumption of the divider. By changing $V_{b,n}$ from 250 to 700 mV and $V_{b,p}$ from 660 to 110 mV, f_{SO} is tuned from 990 MHz to 30 GHz. Meaning that the divider can operate with an input frequency ranging from 4 up to 120 GHz with a power consumption that increases almost linearly from 0.08 to 6.5 mW. Figure 5.15 shows

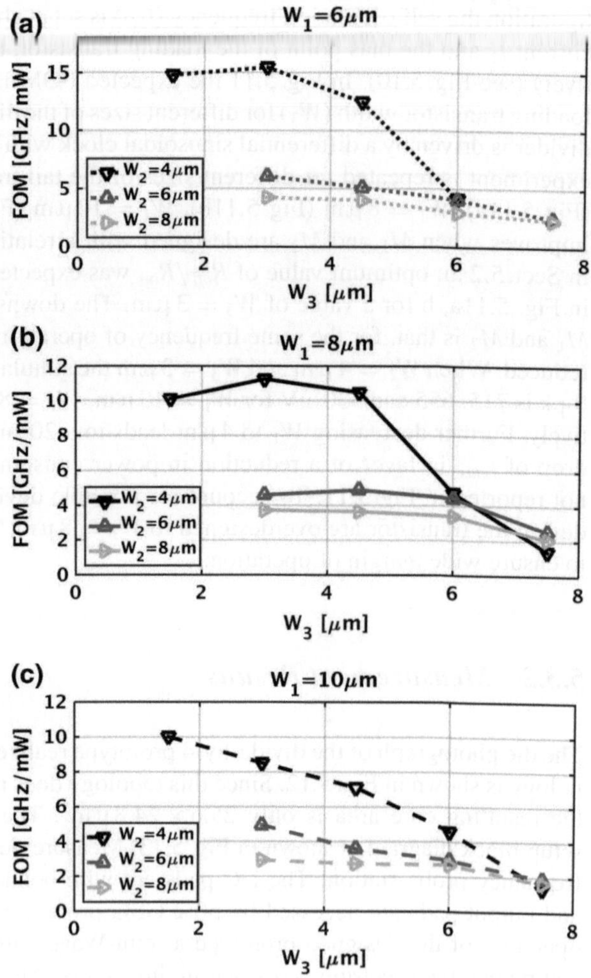

Fig. 5.11 Simulated FOM versus the loading transistor width (W_3) for different sizes of the differential pair (W_2) when the width of the tail transistor is $W_1 = 6\,\mu m$ (**a**), $W_1 = 8\,\mu m$ (**b**), $W_1 = 10\,\mu m$ (**c**). All transistor are minimal length. For fair comparison, the f_{SO} is set at about 80 GHz for each configuration. ©2015 IEEE. Reprinted, with permission, from [21]

the measured sensitivity curves and respective power consumption. The prototype is locked from a 25 to a 102 GHz input clock frequency, demonstrating a minimum and maximum locking range of 4.2 and 42.4% respectively. Noteworthy, the frequency band from 44.3 to 90 GHz is covered in only three steps with a minimum fractional bandwidth in exceed of 20% and power consumption less than 4.7 mW. Further characterization of the sensitivity curves under signal injection is limited by the band-pass behavior of the on-chip balun. In Fig. 5.16 the measured phase noise at the input and output of the divider for an 80 GHz input clock frequency is reported. The expected $20log_{10}(4) \approx$12dB phase noise reduction is demonstrate up to $\approx$5 MHz. The phase noise far from the carrier is limited by the on-chip buffer. However, the noise contribution of the divider is low-pass filtered by the loop when employed in a PLL and

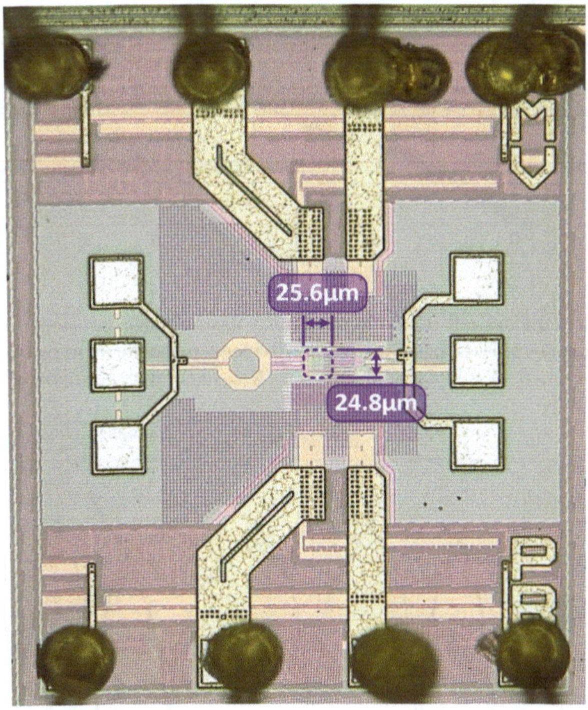

Fig. 5.12 Die micrograph of the realized test chip (core dimensions: 25.6 μm × 24.8 μm). ©2015 IEEE. Reprinted, with permission, from [21]

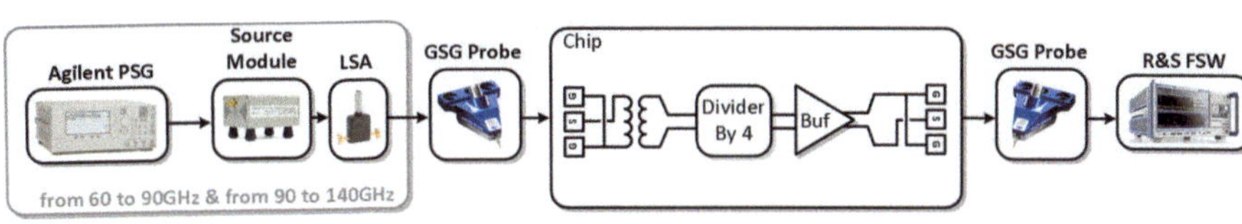

Fig. 5.13 Divider test chip and measurement setup block diagram. ©2015 IEEE. Reprinted, with permission, from [21]

the loop bandwidth of state-of the-art mm-Wave PLLs is normally limited to 1 ∼ 3 MHz [1, 2].

Table 5.1 summarizes the performance and compares this design with state-of-the-art mm-Wave divide-by-4 in bulk CMOS. This work shows the higher tuning capability and the lower area consumption. Moreover, Table 5.1 and Fig. 5.17 clearly show that, benefited by the presented design techniques, (1) the results reported in [9] are extended towards lower power and higher frequencies and (2) the use of on-chip inductors in (*LC*) dividers comes at the cost of large silicon area and poor tuning capabilities, while exacerbating the problem of magnetic coupling to and from other circuits, and with no obvious improvement with technology scaling.

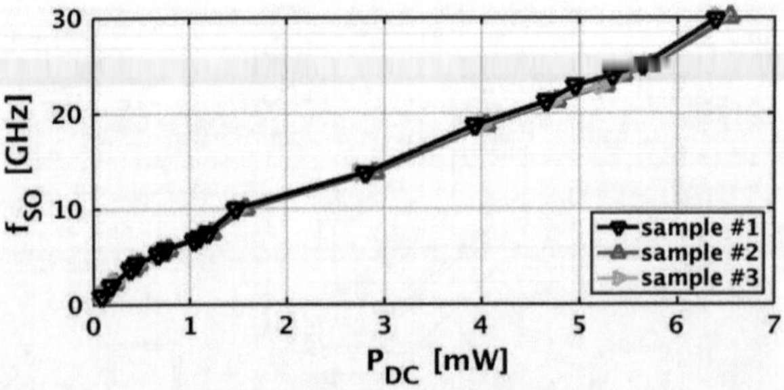

Fig. 5.14 Measured divider self-oscillation frequency against power consumption from three samples. The maximum f_{SO} of 30 GHz shows that the divider can operate up to $f_{IN} = 120$ GHz drawing less than 7 mW from a 0.9 V supply. ©2015 IEEE. Reprinted, with permission, from [21]

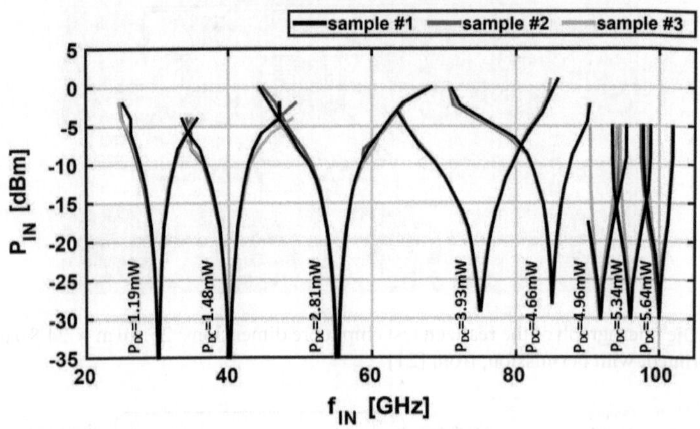

Fig. 5.15 Measured divider sensitivity curve and power consumption from three samples. ©2015 IEEE. Reprinted, with permission, from [21]

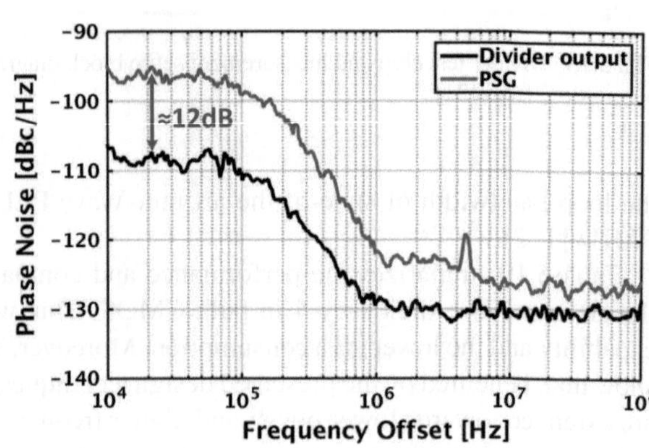

Fig. 5.16 Input (grey) and output (black) measured phase noise from an 80 GHz input frequency. ©2015 IEEE. Reprinted, with permission, from [21]

Table 5.1 State-of-the-Art mm-Wave divide-by-4. ©2015 IEEE. Reprinted, with permission, from [21]

Ref.	f_{min}–f_{max} [GHz]	Locking range [%]	P_{DC} [mW]	FOM[a] [GHz/mW]	Area [mm^2]	CMOS Tech. [nm]
[3]	62.9–71.6	3.2	2.8	0.82	14300	90
[4]	82.5–89	7.6	3	2.17	6380	90
[5]	67–72.4	7.7	15.5	0.35	661200	90
[19]	79.7–81.6	2.4	12	0.16	34980	65
[20]	58.5–72.9	21.9	2.2	6.55	41600	65
[9]	14–70	60–90[b]	1.3–4.8	6.67–17.5[b]	990	32
This work [21]	25–102	4.2–42.4[b]	2.81–5.64	0.74–8.49[b]	635	28

[a]FOM = Locking Range [GHz]/P_{DC} [mW]
[b]Worst - best

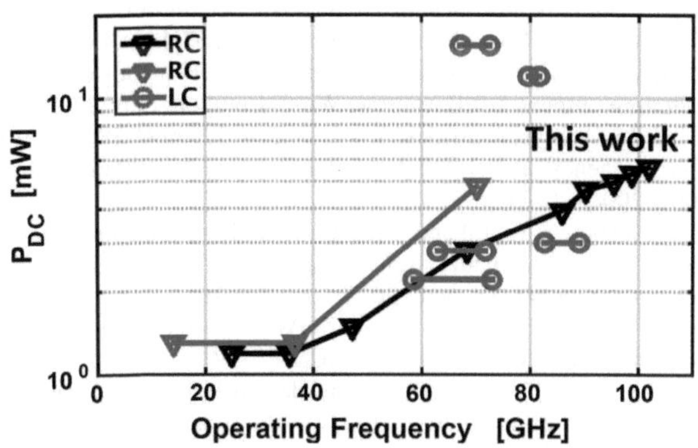

Fig. 5.17 Comparison of mm-Wave CMOS divide-by-4. *RC* based dividers clearly show a superior tuning capability that makes them robust against PVT variation. Moreover, getting rid of the on-chip inductor, they enjoy the full advantage of CMOS scaling (i.e. lower power, lower area and higher speed for each technology node). ©2015 IEEE. Reprinted, with permission, from [21]

5.4 Conclusion

This chapter discussed the fundamentals of high speed frequency dividers. First, the concepts and techniques of injection locking were revised in Sect. 5.1. Then, in Sect. 5.2 the most popular techniques adopted in state-of-the-art high speed frequency dividers were briefly recalled, with the aim of providing design insight into the trade-offs of each topology.

Finally, Sect. 5.3 presented a systematic design methodology to maximize performance of wideband static dividers based on CML dynamic latches. The 25.6 × 24.8 μm^2 28 nm CMOS divide-by-4 prototype shows a measured operating frequency range from 25 to 102 GHz for 5.64 mW maximum power consumption. Measurements repeated on several samples show negligible differences. To the best of the authors knowledge, this is the first time that a single low power divide-by-4 circuit is demonstrated with wide margin over the whole E-Band (60–90 GHz) and beyond.

References

1. X. Yi, C.C. Boon, H. Liu, J.F. Lin, W.M. Lim, A 57.9–68.3 GHz 24.6 mW frequency synthesizer with in-phase injection-coupled QVCO in 65 nm CMOS technology. IEEE J. Solid-State Circuits **49**(2), 347–359 (2014)
2. W. Wu, R.B. Staszewski, J.R. Long, A 56.4–63.4 GHz multi-rate all-digital fractional-N PLL for FMCW radar applications in 65 nm CMOS. IEEE J. Solid-State Circuits **49**(5), 1081–1096 (2014)
3. K. Yamamoto, M. Fujishima, 70 GHz CMOS harmonic injection-locked divider, in *IEEE International Solid State Circuits Conference - Digest of Technical Papers* (San Francisco, CA, 2006), pp. 2472–2481
4. C.C. Chen, H.W. Tsao, H. Wang, Design and analysis of CMOS frequency dividers with wide input locking ranges. IEEE Trans. Microw. Theory Tech. **57**(12), 3060–3069 (2009)
5. C.A. Yu, T.N. Luo, Y.J.E. Chen, A V-band divide-by-four frequency divider with wide locking range and quadrature outputs. IEEE Microw. Wirel. Compon. Lett. **22**(2), 82–84 (2012)
6. L. Wu, H.C. Luong, Analysis and design of a 0.6 V 2.2 mW 58.5–72.9 GHz divide-by-4 injection-locked frequency divider with harmonic boosting. IEEE Trans. Circuits Syst. I Regul. Pap. **60**(8), 2001–2008 (2013)
7. K. Katayama, S. Amakawa, K. Takano, M. Fujishima, Parasitic conscious 54 GHz divide-by-4 injection-locked frequency divider, in *IEEE International Symposium on Radio-Frequency Integration Technology (RFIT)* (Sendai, 2015), pp. 103–105
8. L. Li, P. Reynaert, M. Steyaert, A 60 GHz 15.7 mW static frequency divider in 90 nm CMOS, in *Proceedings of ESSCIRC* (Seville, 2010), pp. 246–249
9. A. Ghilioni, A. Mazzanti, F. Svelto, Analysis and design of mm-wave frequency dividers based on dynamic latches with load modulation. IEEE J. Solid-State Circuits **48**(8), 1842–1850 (2013)
10. B. Razavi, A study of injection locking and pulling in oscillators. IEEE J. Solid-State Circuits **39**(9), 1415–1424 (2004)
11. A. Mirzaei, M.E. Heidari, R. Bagheri, S. Chehrazi, A.A. Abidi, The quadrature LC oscillator: a complete portrait based on injection locking. IEEE J. Solid-State Circuits **42**(9), 1916–1932 (2007)
12. A. Mirzaei, M. Mikhemar, H. Darabi, 21.8 A pulling mitigation technique for direct-conversion transmitters, in *IEEE International Solid-State Circuits Conference Digest of Technical Papers (ISSCC)* (San Francisco, CA, 2014), pp. 374–375
13. W. Sansen, 1.3 analog CMOS from 5 micrometer to 5 nanometer, in *IEEE International Solid-State Circuits Conference - (ISSCC) Digest of Technical Papers* (San Francisco, CA, 2015), pp. 1–6
14. H. Jia, L. Kuang, Z. Wang, B. Chi, A W-band injection-locked frequency doubler based on top-injected coupled resonator. IEEE Trans. Microw. Theory Tech. **64**(1), 210–218 (2016)
15. G. Mangraviti et al., Design and tuning of coupled-LC mm-wave subharmonically injection-locked oscillators. IEEE Trans. Microw. Theory Tech. **63**(7), 2301–2312 (2015)
16. A. Li, S. Zheng, J. Yin, X. Luo, H.C. Luong, A 2148 GHz subharmonic injection-locked fractional-N frequency synthesizer for multiband point-to-point backhaul communications. IEEE J. Solid-State Circuits **49**(8), 1785–1799 (2014)
17. B. Razavi, *RF Microelectronics*, 2nd edn. (Prentice Hall, New Jersey, 2011)
18. D.D. Kim, J. Kim, C. Cho, A 94 GHz locking hysteresis-assisted and tunable CML static divider in 65 nm SOI CMOS, in *IEEE International Solid-State Circuits Conference - Digest of Technical Papers* (San Francisco, CA, 2008), pp. 460–628

19. P. Mayr, C. Weyers, U. Langmann, A 90 GHz 65 nm CMOS injection-locked frequency divider, in *IEEE International Solid-State Circuits Conference. Digest of Technical Papers* (San Francisco, CA, 2007), pp. 198–596
20. L. Wu, H.C. Luong, A 0.6 V 2.2 mW 58–73GHz divide-by-4 injection-locked frequency divider, in *Proceedings of the IEEE 2012 Custom Integrated Circuits Conference* (2012), pp. 1–4
21. M. Vigilante, P. Reynaert, A 25-102 GHz 2.81–5.64 mW tunable divide-by-4 in 28 nm CMOS, in *2015 IEEE Asian Solid-State Circuits Conference (A-SSCC)* (2015), pp. 1–4

Chapter 6
mm-Wave Broadband Downconverters

This chapter is devoted to receiver (RX) front-ends for mm-Wave applications. As discussed in Chap. 1 the RX sensitivity sets a straightforward limit to the achievable link distance. Moreover, to achieve wideband operation without jeopardizing the insertion loss of the required matching networks, the design techniques discussed in Chap. 3 are largely employed. The design challenges and trade-offs relevant to mm-Wave circuits implemented in deep-scaled CMOS are addressed.

The most popular RX architectures are revised in Sect. 6.1. Sections 6.2 and 6.3 discuss the design techniques proper to low-noise amplifiers and downconversion mixers respectively. The design and measurements of an E-Band LNA are presented in Sect. 6.4. A fully integrated sliding-IF receiver for E-Band point-to-point wireless links is discussed in Sect. 6.5. Both designs are implemented in a 28 nm bulk CMOS process without ultra-thick top metal option and excel for bandwidth of operation while achieving state-of-the-art noise figure.

The results of the E-Band LNA in Sect. 6.4 have been published in the International Solid-State Circuits Conference (ISSCC 2016) and the results of the E-Band receiver in Sect. 6.5 have been published in the IEEE Journal of Solid-State Circuits (JSSC 2017, vol. 52, no. 8).

6.1 Receiver Architectures

Figure 6.1 shows four of the most popular receiver architectures. The most conceptually simple and practically difficult one is depicted in Fig. 6.1a. Here an analog-to-digital converter (ADC) samples the signal directly at RF and feeds it to the digital logic in charge of the signal processing. Even if the use of this architecture is not possible at mm-Wave, it has never been really adopted even in the low GHz range.

M. Vigilante and P. Reynaert, *5G and E-Band Communication Circuits in Deep-Scaled CMOS*, Analog Circuits and Signal Processing,
https://doi.org/10.1007/978-3-319-72646-5_6

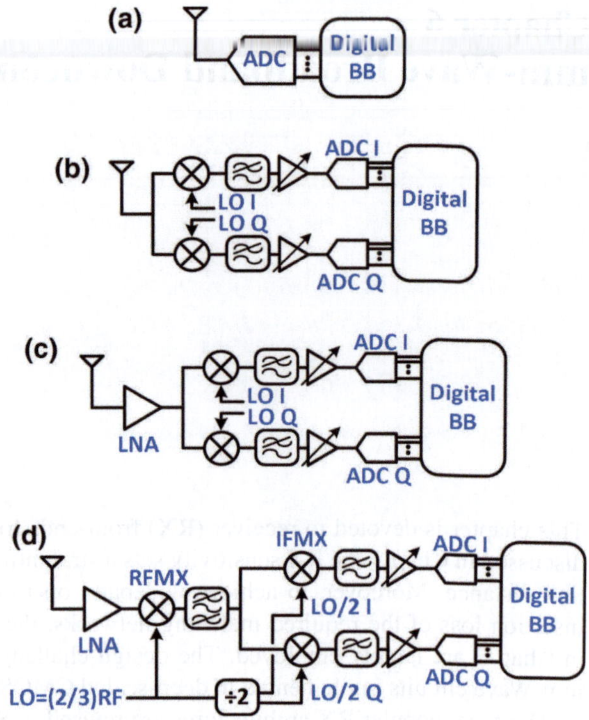

Fig. 6.1 Popular receiver architectures. **a** Software defined radio. **b** Direct-conversion mixer first receiver. **c** Direct-conversion, classical architecture. **d** Two-step downconversion sliding-IF receiver

The reasons are simple, the requirements on speed, linearity and noise on the ADC are so stringent that even if it would be possible to meet them in theory, in practice the power needed would be prohibitively high. Therefore, an analog front-end to filter, downconvert and properly amplify or attenuate the signal prior to the ADC is extremely beneficial.

Figure 6.1b shows a direct-conversion mixer first RX. This topology is state-of-the-art in linearity and largely adopted in the low GHz range [1–3]. However, this topology trades linearity with noise figure. As discussed already in Chap. 1, this trade-off is extremely exacerbated at mm-Wave. Moreover, due to the high free-space path loss and the use of beamforming architectures, the presence of in-band blockers with high power is less of a concern [4]. Nevertheless, some attempts have been demonstrated recently in literature [5].

A classical direct-conversion receiver is shown in Fig. 6.1c. This topology is a state-of-the-art choice for high sensitivity, with a limited impact on linearity both in the low GHz range [2, 3, 6] and at mm-Wave [7–10]. It is worth noting that there is a major difference between voltage mode of operation and current one. The latter approach is more favorable to supply scaling and yield better linearity. However, it is not fully applicable at mm-Wave due to the low gain of single stage amplifiers. The interested reader is referred to [3] for a more in-depth discussion. The direct-conversion architecture has a major drawback though. I/Q LO signals are needed directly at RF. This implies the following. (1) Generating and distributing the LO becomes a key design challenge, both to keep the quadrature error and power consumption under control [11]. (2) As clearly shown in Chap. 4, phase noise and tuning range get worse with frequency. Further challenging the link budget analysis discussed in Chap. 1. (3) LO feed through and DC offset are a major issue [12]. (4) LO and power amplifier (PA) run at the same frequency, causing serious concerns about LO pulling [7, 13–15].

To tackle the aforementioned challenges of direct conversion architectures, a two-step downconversion RX may be used [12]. Figure 6.1d shows an example of sliding-IF RX that needs a single PLL running at 2/3 lower center frequency and tuning range, with no need for distributing quadrature outputs at the carrier frequency. It is worth mentioning that in this case more building blocks are needed. Therefore, power consumption and linearity will suffer. Another problem that worth mentioning is the image rejection. In this case an extra filter is needed to reject the image after the first downconversion mixer (see Fig. 6.1d). This last issue is extremely challenging to solve in the low GHz range [12]. An elegant solution can be found in [16]. However, this is much less of a concern at mm-Wave, where the spectrum of the wanted signal and its image are well spaced in frequency and tuning circuits are anyhow needed to achieve the required gain [17].

6.2 Low-Noise Amplifiers Basics

6.2.1 *Challenges @mm-Wave*

The low-noise amplifier is the first block in the receiver chain, see Fig. 6.1. It needs to provide low noise figure and gain high enough to dominate over the losses and noise of the following blocks. This basic issue is well represented in the Friis' equation [12, 18]

$$NF_{tot} = 1 + (NF_1 - 1) + \frac{NF_2 - 1}{A_{P,1}} + \cdots + \frac{NF_n - 1}{A_{P,1} \cdot \cdots \cdot A_{P,n-1}}, \tag{6.1}$$

where $A_{P,i}$ and NF_i are the power gain and noise figure of the ith stage of the chain.

At the same time the LNA needs to provide a 50 Ω input impedance to interface the antenna. Meaning that an input matching network is needed, and its insertion loss is directly added to the noise figure of the RX. Figure 1.5 clearly shows the trend of power gain and noise figure against frequency of operation for a single stage amplifier implemented in 28 nm CMOS. The end result on state-of-the-art LNAs performance is reported in Fig. 6.2.

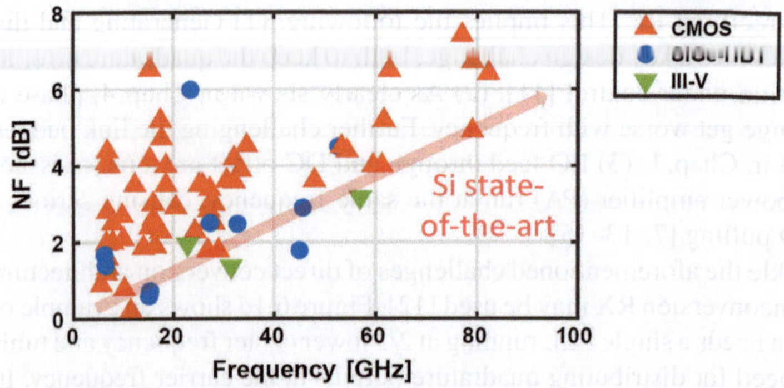

Fig. 6.2 Noise figure against frequency of state-of-the-art low-noise amplifiers [19]. ©2017 John Wiley and Sons. Reprinted, with permission, from [20]

Furthermore, several gain stages need to be cascaded to achieve the required gain as the frequency of operation approaches a fraction of f_{MAX}. This, together with the aggressive supply scaling reported in Fig. 1.6, severely limits the linearity of mm-Wave LNAs.

6.2.2 Most Adopted Circuits

In the following, the most used single-stage amplifiers for mm-Wave LNAs are briefly revised. Crude approximations (i.e. not valid at mm-Wave) are made with the aim of getting intuitive insight on the basic trade-offs of each topology. In particular, we assume (1) input and output are perfectly isolated. Therefore, the input impedance is not sensitive to the load impedance and there is no problem with stability. (2) We assume that the transistor noise is white, proportional to its transconductance, and can be approximated by a current source tied between drain an source terminals $\overline{i_n}^2 = 4kT\gamma g_m$. (3) We consider only the case of perfect input match to 50 Ω and do not discuss the conditions for optimum noise figure (in general and in practice different from the conditions for maximum power gain [21]).

To get insight is it useful to refer to the circuit in Fig. 6.3, where an ideal transconductor with an input impedance Z_{IN} and output noise $\overline{i_{n,o}}^2$ is considered. The noise figure defined in Eq. 1.3 can be calculated in this case as follow [12]

$$NF = \frac{SNR_{IN}}{SNR_O} = 1 + \frac{\overline{i_{n,o}^2}}{4\,k\,T\,R_S\,|\alpha|^2\,g_m^2} = 1 + \frac{\overline{i_{n,o}^2}}{4\,k\,T\,R_S\,G_m^2}, \tag{6.2}$$

where

$$G_m = \left|\frac{I_o}{V_{in}}\right| = |\alpha|\,g_m = \left|\frac{Z_{IN}}{Z_{IN}+R_S}\right| g_m. \tag{6.3}$$

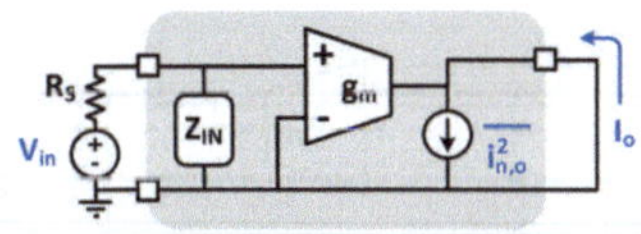

Fig. 6.3 Simplified schematic of a transconductance amplifier

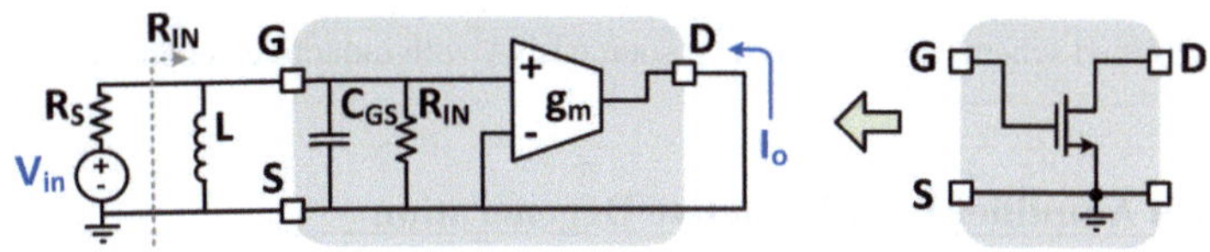

Fig. 6.4 Simplified schematic of a common source LNA

Equation 6.2 shows that the noise figure of this circuit can be calculated as the output noise due to the amplifier divided by the transconductance from V_{in} to I_o (rather than from the transconductor input to its output) normalized by $4\,k\,T\,R_S$ and adding one to this result [12].

Equation 6.3 shows that when $Z_{IN} \rightarrow +\infty$ no power is delivered to the LNA, G_m reaches its maximum equal to g_m and the noise figure in Eq. 6.2 is minimized. However, this condition does not satisfy the input match to R_S.

6.2.2.1 CS Amplifier

The schematic of a single-ended common source amplifier is shown in Fig. 6.4. The capacitance C_{GD} and the output impedance are neglected for simplicity. An inductor L is added in parallel to C_{GS} to resonate at $f_o = 1/(2\pi\sqrt{L\,C_{GS}})$. The transconductance from V_{in} to I_o at f_o is

$$G_m = \left|\frac{I_o}{V_{in}}\right| = \frac{R_{IN}}{R_{IN} + R_S} g_m. \tag{6.4}$$

The resulting noise figure is [12]

$$NF = 1 + \frac{\overline{i_{n,o}^2}}{4\,k\,T\,R_S\,G_m^2} = 1 + \frac{4\,k\,T\,\gamma\,g_m}{4\,k\,T\,R_S\,G_m^2} = 1 + \frac{\gamma(R_{IN} + R_S)}{g_m\,R_S\,R_{IN}}. \tag{6.5}$$

This circuit shows high gain and low noise, but high input impedance. As shown in Chap. 3, a high RC product directly limits the achievable bandwidth. Moreover, there are limits to the impedance transformation ratio that can be practically achieved. This is partially due to the fact that 1:1 transformers shows lower insertion loss, and are always preferred whenever possible [22–24]. We therefore seek a circuit solution that shows a lower input impedance.

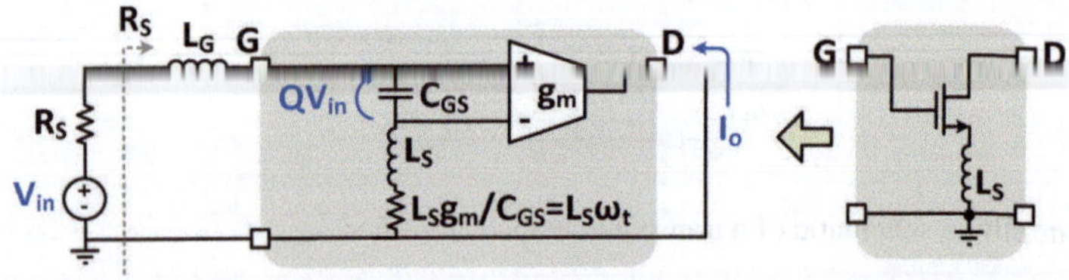

Fig. 6.5 Simplified schematic of a common source LNA with inductive degeneration

6.2.2.2 CS Amplifier with Inductive Degeneration

The schematic of a single-ended common source amplifier with inductive degeneration is shown in Fig. 6.5. The capacitance C_{GD} and the output impedance are neglected for simplicity. The input impedance of this circuit is

$$Z_{IN} = L_S \frac{g_m}{C_{GS}} + s(L_S + L_G) + \frac{1}{s\,C_{GS}}. \tag{6.6}$$

To realize the required 50 Ω input impedance at resonance $\omega_o = 1/(\sqrt{(L_S + L_G)C_{GS}})$ the following condition is imposed

$$R_S = L_S \frac{g_m}{C_{GS}} = L_S\,\omega_t. \tag{6.7}$$

The transconductance from V_{in} to I_o at ω_o can be calculated as [12]

$$G_m = \left|\frac{I_o}{V_{in}}\right| = Q\,g_m = \frac{g_m}{\omega_o\,C_{GS}\,2R_S} = \frac{\omega_t}{\omega_o\,2R_S}. \tag{6.8}$$

While the noise figure is

$$NF = 1 + \frac{\overline{i_{n,o}^2}}{4\,k\,T\,R_S\,G_m^2} = 1 + \frac{k\,T\,\gamma\,g_m}{4\,k\,T\,R_S\,Q^2\,g_m^2} = 1 + \gamma\,g_m\,R_S\left(\frac{\omega_o}{\omega_t}\right)^2. \tag{6.9}$$

Equation 6.8 shows that G_m is inversely proportional to the frequency of operation ω_o, directly resulting in higher noise figure at higher frequencies (see Eq. 6.9). The contrary is true for the cut-off frequency of the transconductor, the higher ω_t the better.

6.2.2.3 CG Amplifier

Another elementary amplifier that shows low input impedance is shown in Fig. 6.6. Here the input and output are assumed isolated. When this is not the case, the picture changes considerably. Especially in deep-scaled CMOS where the channel-length

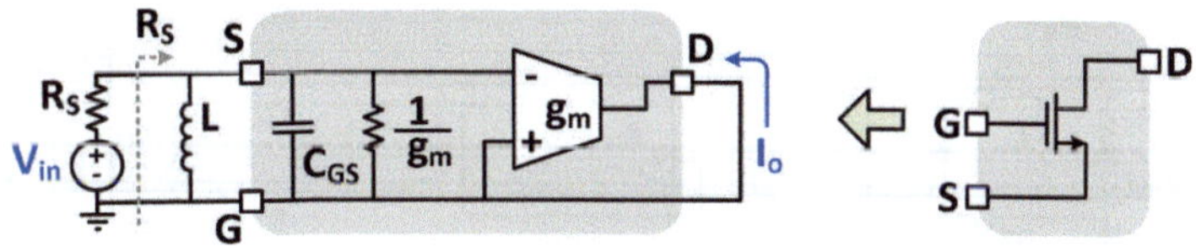

Fig. 6.6 Simplified schematic of a common gate LNA

modulation effect is strongly present [12]. However, under this simplified assumption, the condition of matching at the resonant frequency $f_o = 1/(2\pi\sqrt{L\,C_{GS}})$ is realized when

$$R_S = \frac{1}{g_m}. \tag{6.10}$$

The transconductance from V_{in} to I_o at ω_o under this condition is simply

$$G_m = \left|\frac{I_o}{V_{in}}\right| = \frac{g_m}{2}. \tag{6.11}$$

While the noise figure is [12]

$$NF = 1 + \frac{\overline{i_{n,o}^2}}{4\,k\,T\,R_S\,G_m^2} = 1 + \frac{k\,T\,\gamma\,g_m\,4}{4\,k\,T\,R_S\,g_m^2} = 1 + \gamma. \tag{6.12}$$

One of the major limitations of this topology is that the matching condition in Eq. 6.10 directly limits the gain. However, the RC product in this case is extremely low and the linearity remarkably high.

6.2.2.4 Gm-Boosted CG Amplifier

A major improvement on the classical CG amplifier is shown in Fig. 6.7. Here an ideal noiseless gain stage is added with negative polarity between source and gate. Obviously, practical limitations do exist, i.e. if it would be possible to amplify the signal without adding noise, we would use such amplifier as LNA in the first place. Once again, the input and output are assumed isolated for simplicity. To achieve matching the following relation is imposed at the resonant frequency $f_o = 1/(2\pi\sqrt{L\,C_{GS}})$ [6, 25, 26]

$$R_S = \frac{1}{(1+A)\,g_m}. \tag{6.13}$$

The transconductance from V_{in} to I_o at ω_o under this condition is simply

$$G_m = \left|\frac{I_o}{V_{in}}\right| = \frac{(1+A)g_m}{2}. \tag{6.14}$$

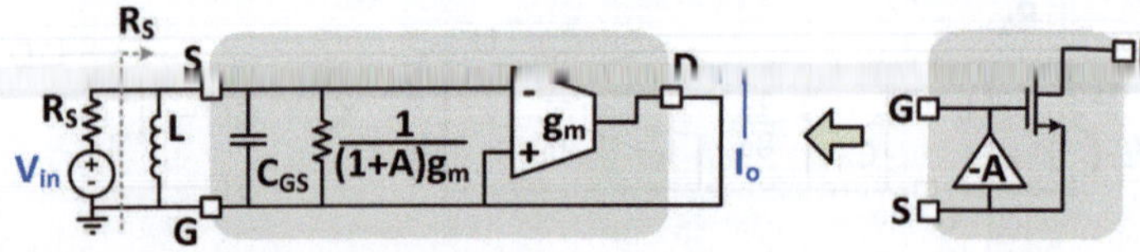

Fig. 6.7 Simplified schematic of a G_m-Boosted common gate LNA

While the noise figure is

$$NF = 1 + \frac{\overline{i_{n,o}^2}}{4kTR_S G_m^2} = 1 + \frac{kT\gamma g_m 4}{4kTR_S(1+A)^2 g_m^2} = 1 + \frac{\gamma}{1+A}. \quad (6.15)$$

This circuit shows interesting performance in terms of gain and noise with a clear improvement on the classical CG topology even if a passive network with a voltage gain $A \approx 1$ is used. Capacitors and transformers may be employed while a cross connection in a differential topology would easily provide the negative sign needed [6, 25, 26].

6.2.3 Cascode Limitations

Leveraging a cascode device is a simple circuit technique that allows to improve reverse isolation and trade bandwidth with gain without a penalty in terms of GBW product or power consumption [12, 27]. This technique would be extremely useful at mm-Wave frequencies, where gain is really needed. However, layout parasitics severely limits the effectiveness of cascode devices at high frequencies.

Figure 6.8a shows the schematic of a CS cascode amplifier and highlights the layout parasitics. An inductor is added to tune out the output capacitance and realize gain at mm-Wave. A parallel resistance models the losses of the inductor at the resonant frequency and constitute the load impedance of the amplifier.

Let us focus on the effect of C_1 first. This capacitance provides a low impedance path between the node X and ground (see Fig. 6.8b), the higher the frequency the lower the impedance. The results are twofold. (1) The leakage current I_{C1} kills the transconductance gain and (2) the noise of the cascode device is no longer negligible [12].

C_2 realizes positive feedback. Due to the Miller effect, C_2 behaves as a negative capacitor ($C_2(1 - A_V)$) in parallel with C_1. Where A_V is the voltage gain between the node X and the output. This effect partially compensate for the losses due to C_1, and it has been proposed to enhance gain of cascode mm-Wave amplifiers [28]. However, positive feedback and negative impedance clearly remind us of oscillators. Indeed, this amplifier resembles the famous Colpitts oscillator in Fig. 6.8c. This

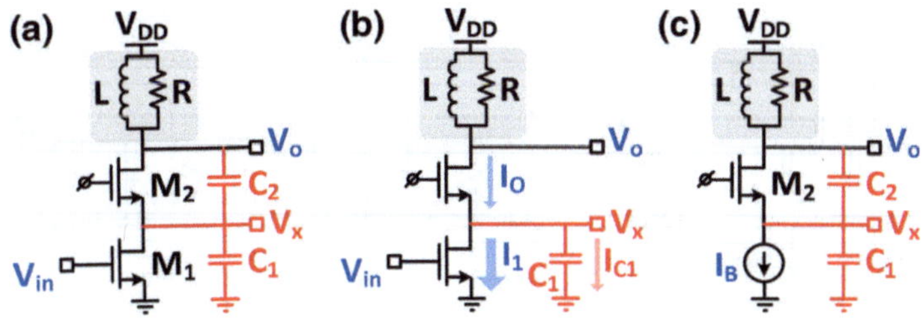

Fig. 6.8 **a** Schematic of a CS cascode amplifier with highlighted layout parasitics. **b** Effect of C_1 at high frequencies. **c** Schematic of a Colpitts oscillator

circuit is prone to oscillate exactly at the resonant frequency f_o, where the amplifier is supposed to operate

$$f_o = \frac{1}{2\pi\sqrt{L\frac{C_1 C_2}{C_1+C_2}}}. \tag{6.16}$$

The start-up condition of this oscillator can be written as [12]

$$g_{m,M2} \geq \frac{(C_1 + C_2)^2}{C_1 C_2 R}. \tag{6.17}$$

Considering the difficulties of modeling layout parasitics and interconnections at mm-Wave, the successful design of this topology is a real challenge. Lastly, it is worth noting that this technique is not favorable to supply scaling (see Fig. 1.6). For this circuit to operate properly, the lower end of the output swing should be larger than the overdrive voltage of two MOS transistors.

6.2.4 *Neutralized CS Amplifier*

Figure 6.9 shows one of the major limitations of a common source amplifier at mm-Wave. C_{GD} is responsible for reverse isolation and stability degradation, greatly challenging the design and modeling of mm-Wave amplifiers. To address this

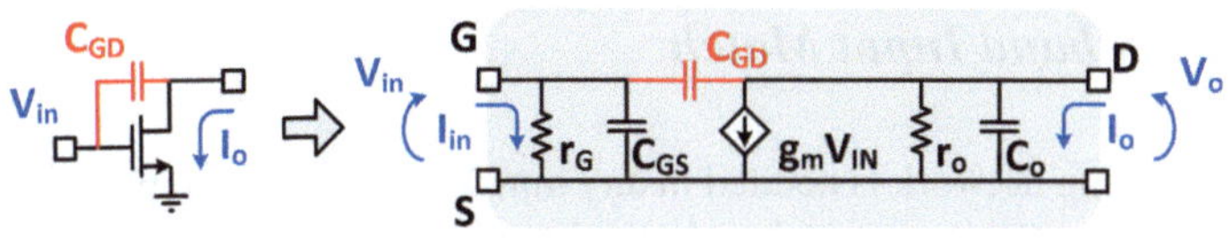

Fig. 6.9 Schematic of a CS single-ended amplifier and its AC model with highlighted the gate-to-drain capacitance

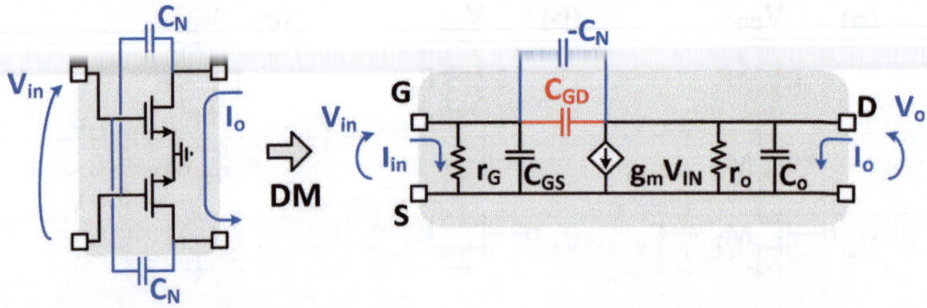

Fig. 6.10 Schematic of a neutralized CS amplifier and its AC equivalent circuit in differential mode

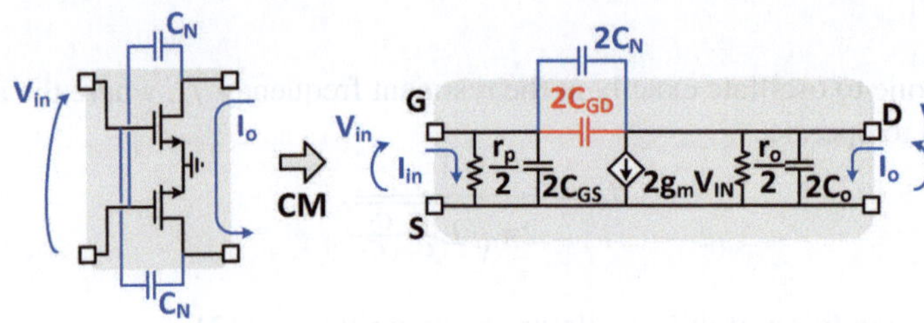

Fig. 6.11 Schematic of a neutralized CS amplifier and its AC equivalent circuit in common mode

limitation, we speculate that a negative capacitor of the same value in parallel with C_{GD} would solve the problem.

This is indeed the case and a differential implementation would provide a simple and extremely effective solution, see Fig. 6.10 [21, 22]. This circuit enables unconditional stability in differential mode, high reverse isolation and high gain at mm-Wave and is widely adopted whenever a differential amplifier is needed.

However, this circuit does also have drawbacks. The most important one is shown in Fig. 6.11. For a common mode (CM) signal the problem is not only still present, but 2 times C_N worse. Great care should be taken when the bias lines are designed. Large bias resistors are normally added in the low current common mode path [22]. Moreover, as shown in Chap. 4 the second harmonic current flows in common mode. Realizing a proper control of the CM impedance is extremely challenging also because of the lack of reverse isolation. We will return on this point when discussing power amplifiers.

6.2.5 *Broadband Input Match*

An input matching network is needed in any mm-Wave receiver to resonate the input capacitance of the first stage and pads, and realize impedance scaling to interface the antenna (normally 50 Ω). When a differential implementation is used, a balun is also needed. For all these reasons, the design techniques discussed in Chap. 3 are fundamental.

Here, we specifically focus on the input match and the effect of the RC product of the input impedance of the LNA on the achievable bandwidth. The absolute value of S_{11} is a good measure of input matching and is widely adopted in literature. When $|S_{11}| < -10\,\text{dB}$ more than 90% of the power is delivered to the load and the amplifier is considered matched [12]. Some margin is needed to account for model inaccuracy. It is worth noting that when a $50\,\Omega$ voltage source is terminated on a $95\,\Omega$ real impedance, it results in $|S_{11}| \approx -10.16\,\text{dB}$ and the condition of matching is already achieved. Therefore, perfect matching to $50\,\Omega$ is not needed.

To put this discussion in prospective, we design an input matching network to realize $|S_{11}| < -10\,\text{dB}$ over a wide range of frequencies around $f_o = 80\,\text{GHz}$. As already done in Chap. 3, $\omega_L = 2\pi\,68\,\text{GHz}$ and $\omega_H = 2\pi\,92\,\text{GHz}$ are imposed. Further, $C_{in} = 14\,fF$ is considered and the RC product of the load is progressively lowered by acting on the value of R_{in}. First the filter with no impedance scaling in Fig. 6.12a is designed and $S_{11} = (Z_{in} - R_{in})/(Z_{in} + R_{in})$ is evaluated. When the desired frequency response is achieved, the circuit transformation in Fig. 6.12b is applied. This results in n times lower input impedance and the same S_{11}.

The result of this investigation is shown in Fig. 6.13. Clearly, for a given bandwidth and ripple, the quality factor of the load needs to be chosen low enough. This is the reason why it is not possible to realize a broadband input match for a common source amplifier with a simple low-loss 4th order filter. Moreover, if the quality factor drops too much, the bandwidth needs to be increased, otherwise the matching condition can not be achieved.

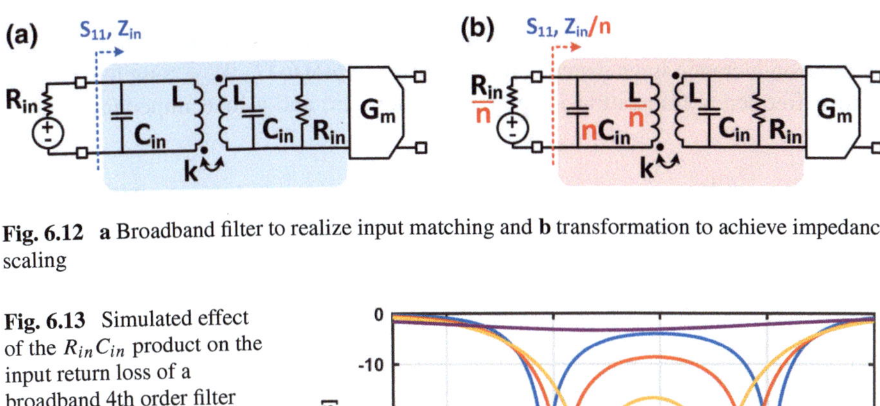

Fig. 6.12 **a** Broadband filter to realize input matching and **b** transformation to achieve impedance scaling

Fig. 6.13 Simulated effect of the $R_{in}C_{in}$ product on the input return loss of a broadband 4th order filter

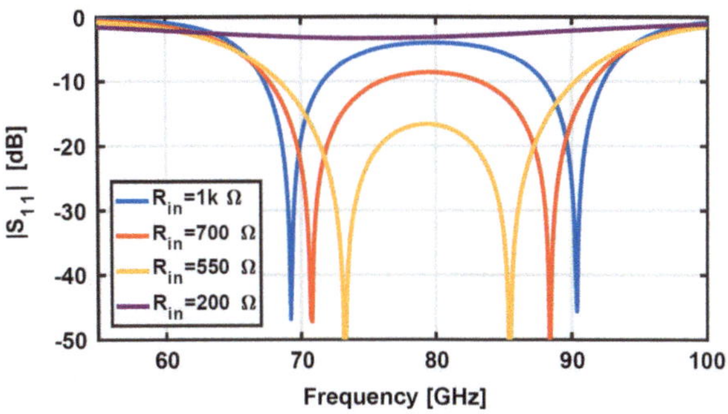

In addition, even if the LNA shows an input impedance of $C_{in} = 14fF$ and $R_{in} = 550\,\Omega$ making it theoretically possible to realize the desired matching shown in Fig. 6.13, still the required impedance transformation is remarkably high $n = R_{in}/50 = 11$. This implies that the primary inductor should be designed 11 times smaller than the secondary (which might be not feasible) and the parallel capacitor at the input 11 times larger than C_{in} (which may lead to unacceptable losses). Finally, we have so far neglected the losses of the matching network. In a real implementation those are not negligible. For all these reasons the first stage of the LNA should present a low resistive impedance and the design of both LNA and its matching network presents non straightforward trade-offs.

6.3 Downconversion Mixers @mm-Wave

A downconversion mixer is needed to realize frequency translation and feed the received signal to the baseband circuitry. To perform this task, a non-linear time-variant circuit is needed, making its theoretical discussion not trivial. Figure 6.14a shows one of the most popular active mixer circuit for mm-Wave applications. When compared to a single-balanced implementation, a double-balanced one permits, among other advantages, to greatly suppress the LO and RF feedthrough, while enhancing the IP_2 [12]. This is particularly critical, since the second order intermodulation distortion falls at low frequencies, where the downconverted signal is to be received.

The conceptual block diagram of a downconversion mixer is shown in Fig. 6.14b. A transconductance stage generate the RF current equal to $G_m\ V_{RF}$, a switching-quad realize frequency translation in the current domain, and a load impedance realize

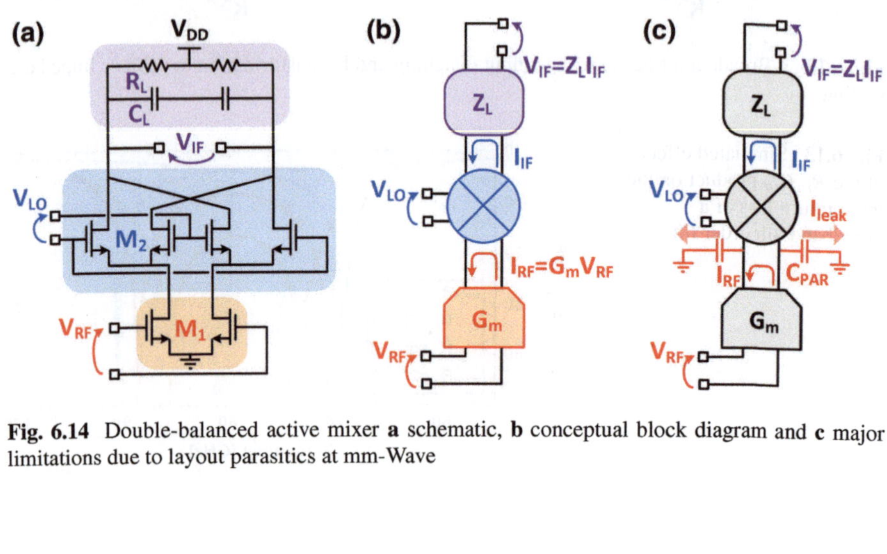

Fig. 6.14 Double-balanced active mixer **a** schematic, **b** conceptual block diagram and **c** major limitations due to layout parasitics at mm-Wave

current-to-voltage conversion and low-pass filtering. Assuming a square wave LO with 50% duty cycle the IF current and voltage can be expressed as [12]

$$I_{IF} \approx \frac{2}{\pi} G_m V_{RF} \{cos[(\omega_{RF} - \omega_{RF})t] + cos[(\omega_{RF} + \omega_{RF})t] + \ldots\}, \tag{6.18}$$

$$V_{IF} \approx \frac{2}{\pi} G_m V_{RF} R_L cos[(\omega_{RF} - \omega_{RF})t]. \tag{6.19}$$

The resulting conversion gain (CG) is

$$CG = \frac{V_{IF}}{V_{RF}} \approx \frac{2}{\pi} G_m R_L. \tag{6.20}$$

Equation 6.20 shows an evident trade-off between CG and linearity. The higher R_L, the higher the CG. However, a large voltage swing at the output severely limits linearity, especially at low supply voltage. Moreover, the linearity is already challenged by the device stacking, similarly to what already discussed for cascode devices. For these reasons current-mode receivers were successfully proposed as high-linearity alternative for low voltage applications. In these case, the transconductance stage is AC coupled to a passive switching-quad allowing separate DC bias and lower noise. Then, a transimpedance amplifier with low input impedance converts the IF current to voltage [2, 3].

When a downconversion mixer is designed for mm-Wave application, it shows several major challenges. (1) The LO is not a square wave, immediately lowering the CG. (2) C_{PAR} shunts part of the RF current to ground further lowering the CG and increasing noise. (3) The G_m stage needs a relatively large biasing current to provide the necessary f_t. However, a large bias current increases the noise contribution of the switching-quad [12]. Therefore, several state-of-the-art solutions adopt a large number of inductors to resonate layout parasitic and/or degenerate the transconductance stage to improve gain noise and/or linearity; transformers are used to allow separate biasing and 4th order filters are used as load impedance to enhance the GBW product of the first downconversion mixer in sliding-IF receivers [11, 17, 29–31].

6.4 Design Example 1: A Wideband LNA in 28nm CMOS

6.4.1 LNA Architecture

Figure 6.15 shows the block diagram of the proposed E-Band LNA. To limit the detrimental effect on the noise figure of the following blocks, the LNA employs four active stages to realize >30 dB gain while driving a 50 Ω load. A transformer at the input realizes the required single-ended to differential conversion while providing protection against ESD events. A G_m-boosted common gate (CG) amplifier realizes

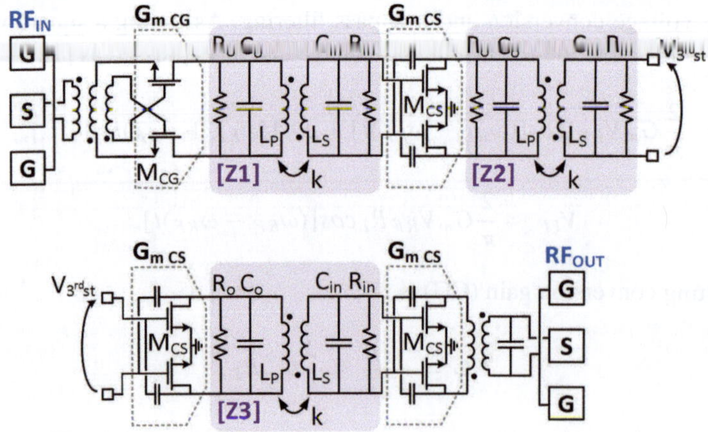

Fig. 6.15 Simplified schematic of the 4-stage LNA test chip. ©2017 IEEE. Reprinted, with permission, from [33]

the required broadband input match to 50 Ω [6, 25, 26]. The other transconductor stages are implemented with neutralized common source (CS) amplifiers for maximum gain and reverse isolation [22, 32]. 4th order inter-stage matching networks based on transformers are used to enhance the GBW product in the RF band. The transistors are designed with $W_{CG} = 35.7\,\mu\text{m}$, $W_{CS} = 25.1\,\mu\text{m}$ and minimum length of 28 nm.

Figure 6.16 shows the main transistor parameters against inversion coefficient. As expected from the theoretical discussion about technology scaling in Sect. 2.1, a clear optimum biasing point is evident for a transistor implemented in 28 nm CMOS. Therefore, all transconductors are biased with an inversion coefficient $IC \approx 1$ for maximum $f_t\, g_m/I_{DS}$ product, resulting in an optimal design for speed and noise for a given power consumption [34, 35]. The transistor layout proposed in [22] is leveraged to minimize the parasitics due to the low level metal interconnects and maximize f_{MAX}. It is worth noting that f_{MAX} and NF_{min} are much more sensitive to layout parasitics than g_m and f_t. Thus, the effect of different finger width for a given W and g_m is better captured by the former. Moreover, mm-Wave transistor models were not available at design time. So the simulated values of f_{MAX} and NF_{min} should be taken with a grain of salt. I.e. the simulated value is expected to be optimistic and the optimum finger width shown in simulations may differ from the measured one.

Figure 6.17 shows the layout view of the realized prototype. The metal length and width are adopted as design variables to set the required inductance value, while magnetic coupling coefficient is set by the spacing between L_P and L_S (a larger distance results in a lower magnetic coupling). From EM simulations, the values of the parasitic magnetic coupling coefficients are estimated as $|k_{p1}| \approx 0.032$ and $|k_{p2}| \approx 0.005$. The effect of different signs of k_{p1} and k_{p2} are shown in Fig. 6.17 (bottom), together with the resulting layout implementations. When the interconnection [Z3A] is adopted (i.e. $k_{p2} > 0$), simulations predict a GBW product above 1 THz with a

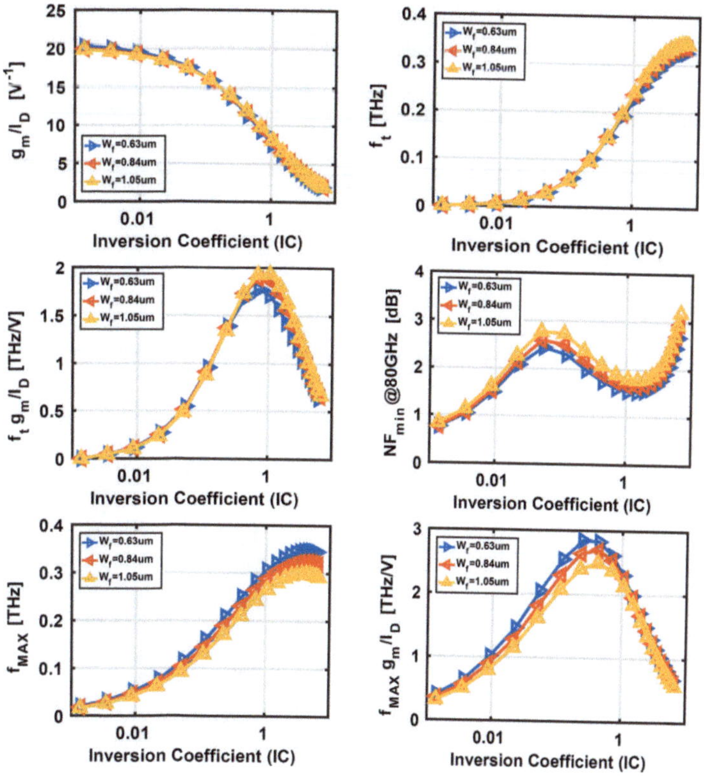

Fig. 6.16 Main transistor parameters versus inversion coefficient for a single transistor CS amplifier implemented in 28 nm CMOS (simulation). W is kept constant to $\approx$30 μm, while the finger width W_f and the number of fingers is varied

ripple of 1.1 dB. Whereas, when the interconnection [Z3B] is used, k_{p2} is negative, resulting in a higher peak gain at the expenses of much larger in-band ripple. The GBW product in this second case drops below 0.5 THz with a ripple larger than 6 dB. To further probe the robustness of these design techniques, 2000 Monte Carlo simulations were performed under both process and mismatch variations. At 3σ the variation on the main design specifications are the following, ±0.7 dB gain, ±1.7 GHz $BW_{-3\,dB}$ and ±0.03 dB noise figure.

6.4.2 *Measurement Results*

Figure 6.18 shows the die micrograph of the realized stand-alone 28 nm bulk CMOS E-Band LNA test chip. The core area is 893 μm × 285 μm, including the input and output RF probe pads. The supply voltage is 0.9V.

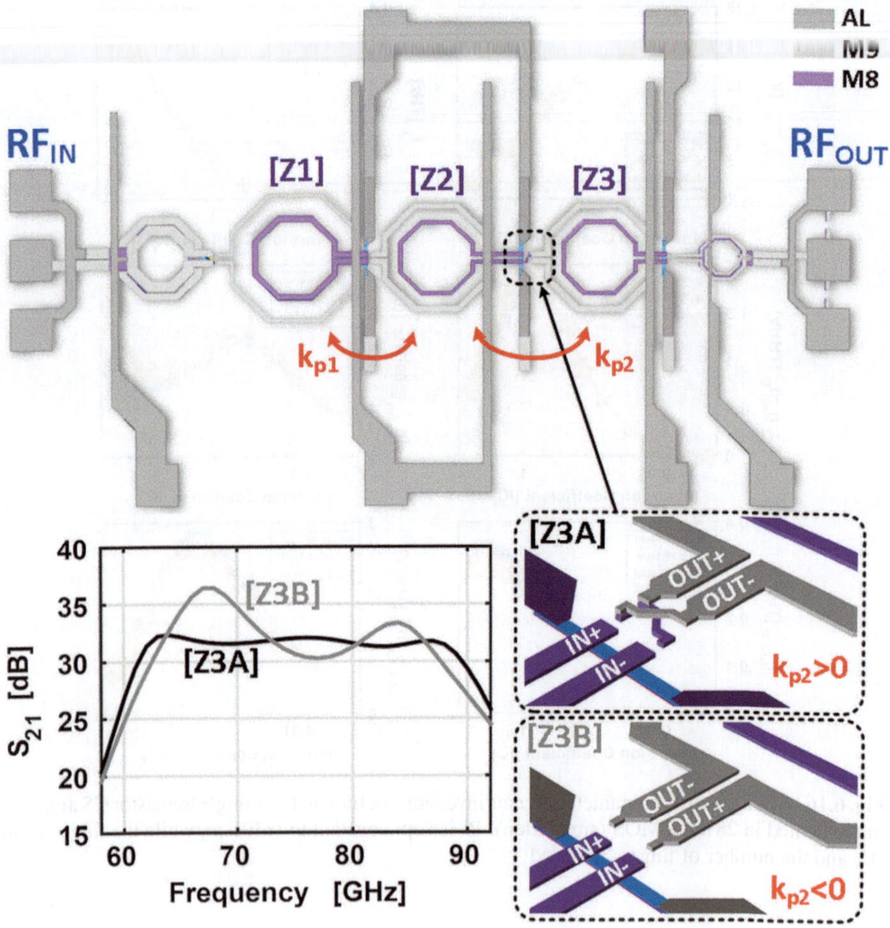

Fig. 6.17 Simulated effect on the gain bandwidth of interconnections Z3A (black line) and Z3B (gray line) that realize different signs of k_{p2}. ©2017 IEEE. Reprinted, with permission, from [33]

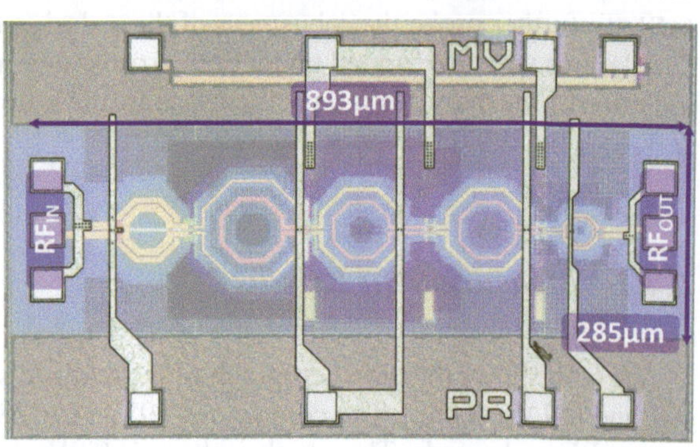

Fig. 6.18 Die micrograph of the 4-stage LNA test chip (core area: 893 µm × 285 µm). ©2017 IEEE. Reprinted, with permission, from [33]

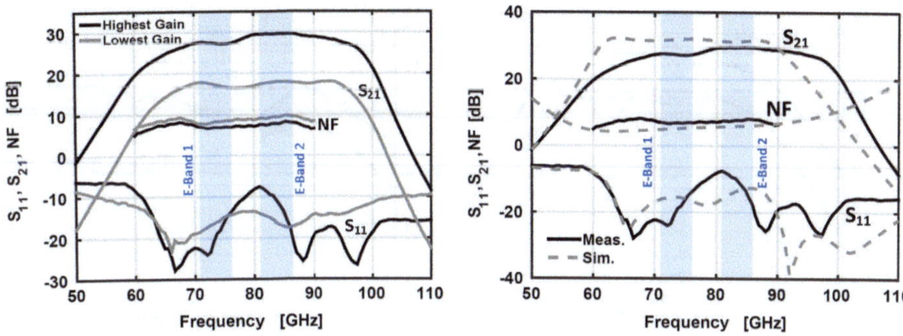

Fig. 6.19 Measured LNA gain, noise figure and input match versus frequency, highest gain (black line) and lowest gain (gray line), left. Measurements against simulation, right. ©2017 IEEE. Reprinted, with permission, from [33]

The measured gain, noise figure and input match at the highest and lowest gain are reported in Fig. 6.19. The measured peak gain is 29.6 dB at 84.1 GHz. The $BW_{-3\,\mathrm{dB}}$ spans from 68.1 to 96.4 GHz at the highest gain, resulting in a GBW product of 0.85 THz. The S_{11} is below −7.6 dB from 59.4 to 110 GHz, showing a broadband input match. The measured gain can be varied from 18 to 29.6 dB while keeping a bandwidth in excess of 28 GHz, by increasing the bias current from 13 to 34.8 mA. The noise figure is evaluated using a SAGE STZ-12-I1 E-Band noise source and a Rohde & Schwarz spectrum analyzer. The measured NF reaches the in-band minimum of 6.4 dB at 89.5 GHz, and varies by less than 2 dB from 68.1 to 90 GHz. The measured worst case group delay varies less than ±21.7ps from 60 to 100 GHz and less than ±12.6 ps in each sub-band. Such a flat group delay in combination with the broadband S_{21} response is key to enable wireless data links with wide modulation bandwidth without deteriorating the EVM [36, 37], proving the effectiveness of the proposed design techniques for broadband operation. Figure 6.19 reports also measurements against simulations. An expected reduction in gain and increase of noise is observed. Moreover, the LNA frequency response is shifted to ≈5 GHz higher frequencies. Benefited by the adopted broadband design techniques, this effect is visible, but has no consequences on the performance over the target bandwidth. It is worth noting that at mm-Wave is not trivial to accurately model all passives and active devices, therefore some differences between measurements and simulation are to be expected and design techniques that prove robust against model inaccuracy are more than welcome. To further prove the robustness of the measurement set-up, several samples of the same LNA are measured. The results of this investigation is shown in Fig. 6.20. A good repeatability of small signal and noise figure measurements is verified.

The large signal continuous wave (CW) measurements at 75 GHz are reported in Fig. 6.21. The same measurements repeated over frequency are shown in Fig. 6.22. The worst case in-band input-referred compression point is −28.1 dBm/−12.3 dBm at the highest/lowest gain.

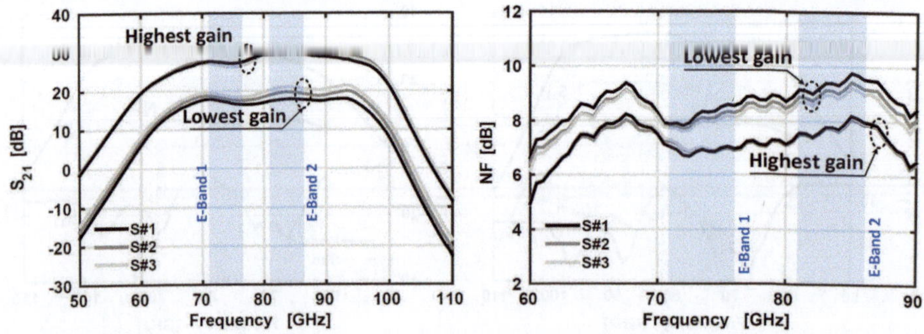

Fig. 6.20 Measured LNA gain (left) and noise figure (right) for 3 different samples (from the same wafer lot)

Fig. 6.21 Measured LNA output power versus input power at 75 GHz, highest gain (black line) and lowest gain (gray line). ©2017 IEEE. Reprinted, with permission, from [33]

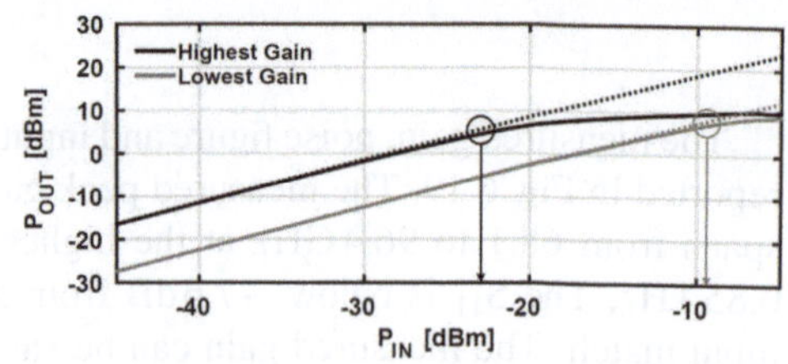

Fig. 6.22 Measured LNA input-referred 1 dB compression point versus frequency, highest gain (black line) and lowest gain (gray line). ©2017 IEEE. Reprinted, with permission, from [33]

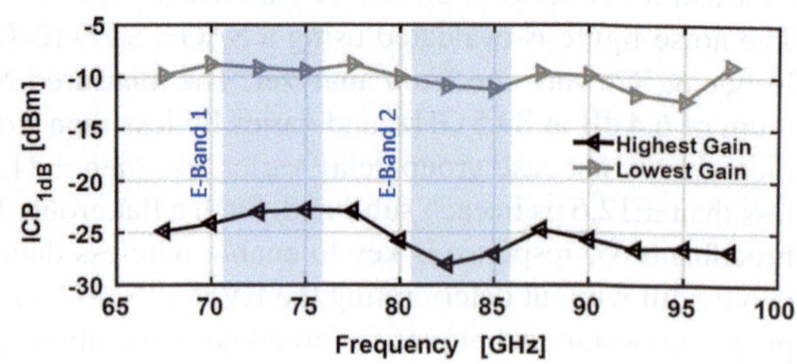

Finally, Table 6.1 summarizes the measured results and compares them with state-of-the-art 70/80 GHz CMOS LNAs. To compare designs at different frequencies the following figure of merit is used [38] in Fig. 6.23

$$FOM = 20log_{10}\left(\frac{Gain_{[lin]} \cdot BW_{[GHz]}}{P_{DC[mW]} \cdot (NF_{lin} - 1)}\right). \tag{6.21}$$

Leveraging the proposed design methodologies, the E-Band LNA achieves a figure of merit ≈10.5 dB better that state-of-the-art designs in the same band and comparable to LNAs at lower frequencies.

Table 6.1 Comparison with state-of-the-art LNAs in 70/80 GHz bands. ©2017 IEEE. Reprinted, with permission, from [33]

Ref.	**This work** [33]		[39]		[40]	[41]
CMOS Tech. (nm)	**28**		28		65	65
$\mathbf{V_{DD}}$ **(V)**	**0.9**		0.9		1.2	1
Gain (dB)	**29.6**	**18**	23.8	19.3	17.5	9.4
$\mathbf{f_c}$ **(GHz)**	**82.3**	**81.1**	79		77	79
$\mathbf{BW_{-3dB}}$ **(GHz)**	**28.3**	**30.7**	10		2	15
GBW (THz)	**0.85**	**0.24**	0.15	0.09	0.01	0.04
NF (dB)	**6.4-8.2**♯	**7.8-9.8**♯	4.9	5.6	7.4	6.7
$\mathbf{ICP_{1dB}}$ **(dBm)**	**−28.1**	**−12.3**	−18.5	−15	−22	n.a.
$\mathbf{P_{DC}}$ **(mW)**	**31.3**	**11.7**	30.6		30	9.7
FOM (dB)	**18.2**	**12.3**	7.7	3.2	−19	1.9

♯In-band noise figure measurements limited by the available noise source to 90 GHz

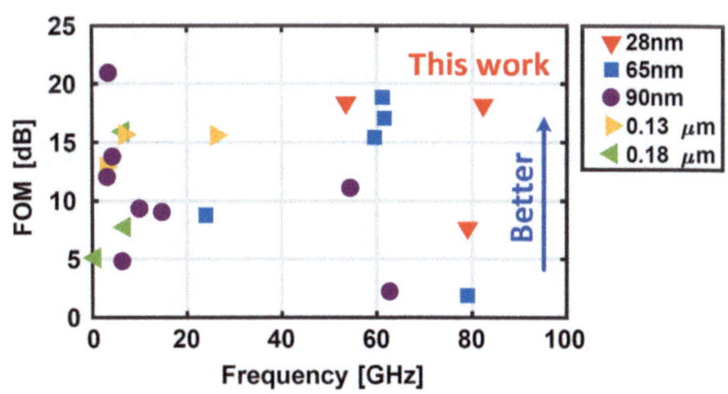

Fig. 6.23 Comparison of recently published state-of-the-art CMOS LNAs. ©2017 IEEE. Reprinted, with permission, from [33]

6.5 Design Example 2: A Wideband Downconverter Front-End in 28nm CMOS

6.5.1 Receiver Architecture

Future 5th generation (5G) mobile networks will deal with data-rate 100× times higher than today [42]. Therefore, E-Band back-haul links are an attractive solution for low-cost fiber extension or replacement over shot to medium distance [43], motivating an increasing research interest in the last few years [15, 44, 45]. However, to realize the full potential of such links a wideband receiver (RX) with high sensitivity and uniform performance over two bands from 71 to 76 GHz and 81 to 86 GHz is needed. Requirements are even more stringent when the spread due to process variations and modeling inaccuracy is considered. Recently, two separate narrowband SiGe BiCMOS receivers targeting either the low or the high band have been reported in [45]. While a narrowband CMOS solution for 79 GHz car radar has

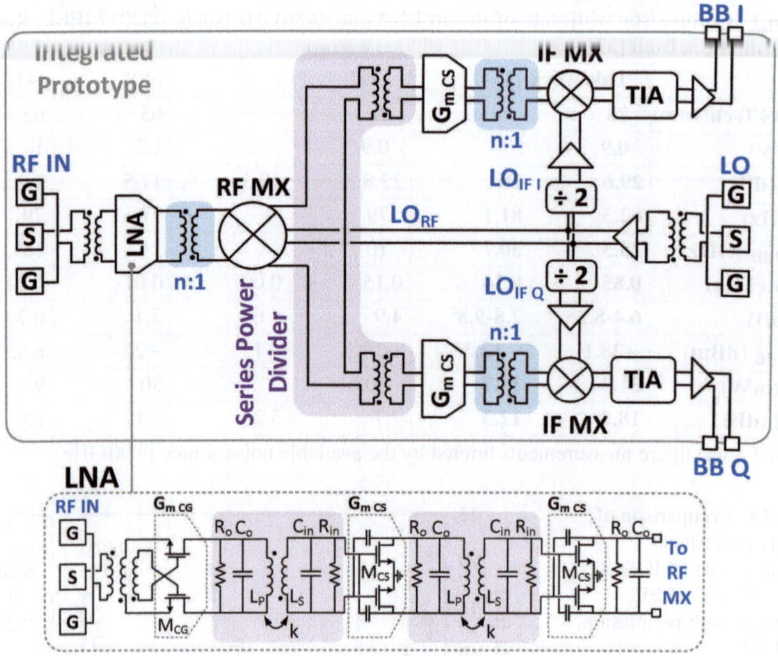

Fig. 6.24 Block diagram of the integrated receiver. ©2017 IEEE. Reprinted, with permission, from [33]

been demonstrated in [46]. However, to the best of the authors' knowledge a single chip CMOS broadband solution has not been reported in literature so far.

The RX adopts a sliding-IF architecture (see Fig. 6.24), allowing to relax the requirements on the local oscillator generation and distribution network, i.e. 2/3 lower center frequency and tuning range, with no need for distributing quadrature outputs at the carrier frequency. However, an additional broadband IF stage is required. This receiver employs a transformer-based series power divider derived from a 4th order 2-port filter to realize >9 GHz (>36%) IF bandwidth. The measured conversion gain is 30.8 dB with <1 dB in-band ripple, the minimum noise figure is 7.3 dB and varies less than 2 dB from 61.4 to 88.9 GHz.

6.5.2 RF Mixer and Power Splitter

The last stage of the LNA drives the RX mixer (implemented with a Gilbert-type switching quad) through a n:1 transformer, Fig. 6.24. This allows a separate DC biasing of the transconductor and the switching quad, yielding better linearity at low V_{DD} [10, 11, 29, 47]. Moreover, the n:1 turns ratio reduces the voltage swing and realizes passive current gain at the secondary.

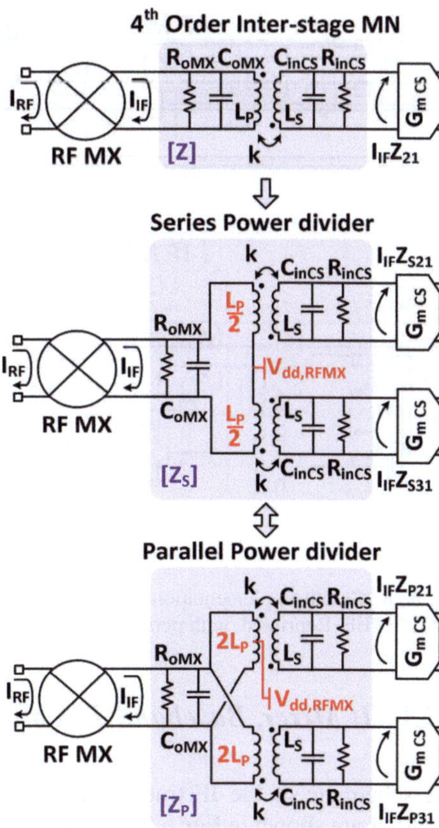

Fig. 6.25 Extension of broadband 4th order 2-port networks based on transformers to realize 3-port series or parallel power dividers. ©2017 IEEE. Reprinted, with permission, from [33]

Figure 6.25 shows two possible implementations of a transformer-based 3-port power divider derived from a 2-port 4th order filter. This transformation does not affect the frequency response, except for a 3 dB magnitude reduction due to ideal power division. Noteworthy, the series power divider results in 4 times less inductance at the primary winding while keeping a symmetrical bias network, leading to lower losses and better common mode rejection when compared with its parallel counterpart. Therefore, the former is preferred in this design.

EM simulations predict a Q-factor of $\approx$6 for a 2nH on-chip inductor at 25 GHz, about 3$\times$ lower than the Q of the inductors used in the LNA inter-stage matching networks at 80 GHz. As discussed in Chap. 3, such a low Q impairs the magnitude of the filter transimpedance at the highest resonant frequency. Therefore, ξ in Eqs. 3.49 and 3.50 is designed lower than 1 to realize the required pre-emphasis. Finally, it is worth noting that the simple transformation to realize a power splitter shown in Fig. 6.25, would not be trivial if realized with a different 4th order network (i.e. Figs. 3.5, 3.6 and 3.8).

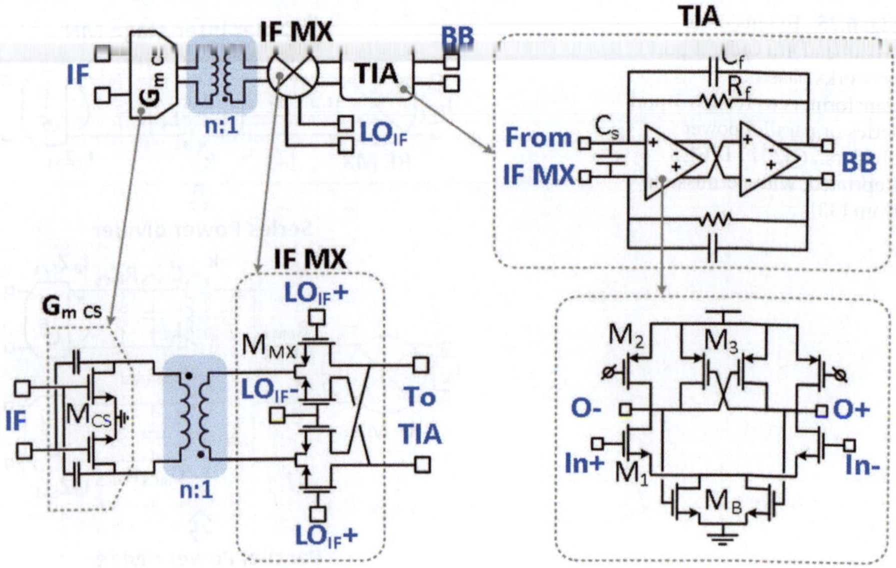

Fig. 6.26 Circuit implementation of the IF transconductor, IF passive mixer and baseband TIA. ©2017 IEEE. Reprinted, with permission, from [33]

6.5.3 If Mixer, Baseband TIA and I/Q Generation

The schematic of the IF transconductor, IF mixer and baseband transimpedance amplifier are shown in Fig. 6.26. A current mode operation is preferred to maximize the linearity of this last stage, bottleneck of the RX chain [6, 10]. A n:1 transformer interfaces the transconductor and the switching quad as in the RF path. The baseband TIA is implemented with a two-stage differential OTA with resistive feedback and provides ≈50 Ω input impedance. The TIA is followed by on-chip buffers to drive the 50 Ω measurement equipment.

The I/Q LO IF signals are generated on-chip by two inductor-less dividers based on CML static latches locked by an external source. Static dividers are preferred against injection locked ones for their larger locking range and lower silicon area, at expenses of higher power consumption [12, 48].

6.5.4 Measurement Results

The receiver prototype is integrated in 28 nm bulk CMOS technology without RF thick top metal option (see Fig. 6.27). GSG probes provide the mm-Wave signals, while the baseband I/Q signals and DC pads are wire-bonded to a PCB. The front-end

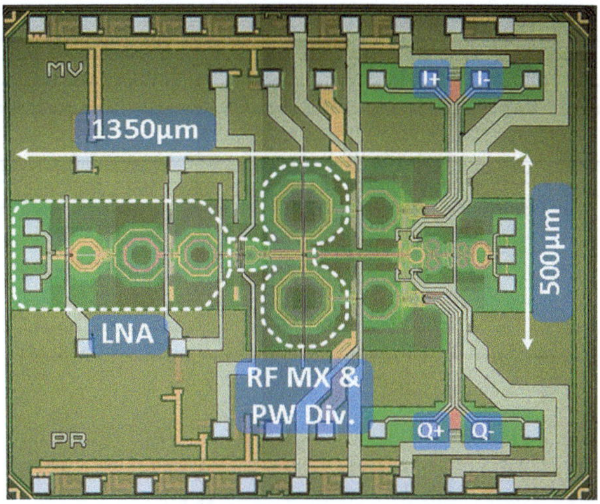

Fig. 6.27 Die micrograph of the receiver test chip. The core area is 1350 µm × 500 µm. The total pad limited area is 1500 µm × 1300 µm. ©2017 IEEE. Reprinted, with permission, from [33]

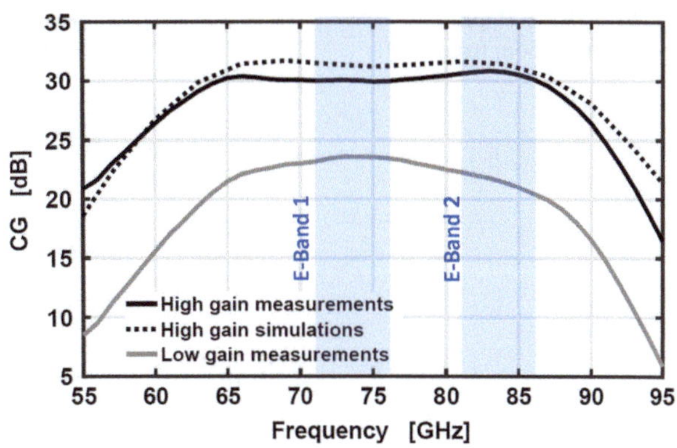

Fig. 6.28 Measured (continuous line) and simulated (dotted line) RX conversion gain versus frequency, high gain (black line) and low gain (gray line). ©2017 IEEE. Reprinted, with permission, from [33]

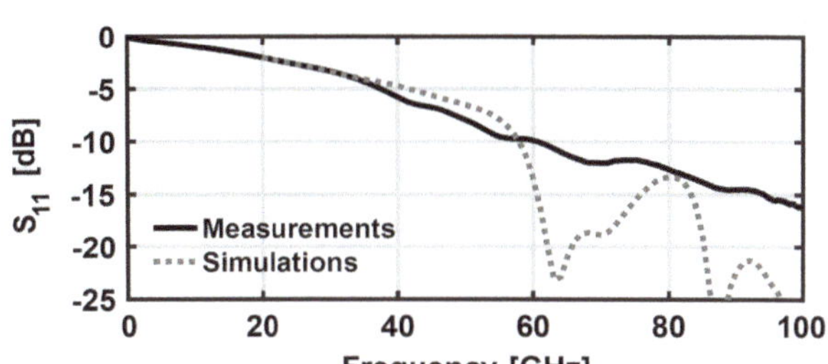

Fig. 6.29 Measured (continuous line) and simulated (dotted line) RX S_{11} versus frequency. ©2017 IEEE. Reprinted, with permission, from [33]

draws 63mA, including the baseband TIAs. The dividers and the buffers driving the IF mixers draw 35mA.

Figure 6.28 shows the measured conversion gain. The measured input match against frequency is reported in Fig. 6.29, while the noise figure is shown in Fig. 6.30. Simulation results are also reported for comparison in dotted line. The RX peak

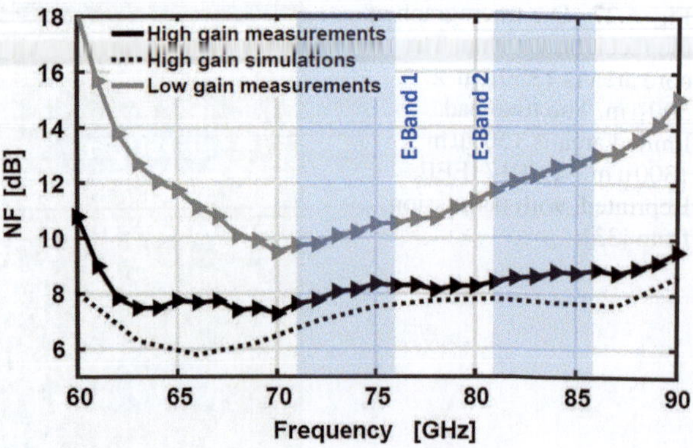

Fig. 6.30 Measured (continuous line) and simulated (dotted line) RX noise figure versus frequency, high gain (black line) and low gain (gray line). ©2017 IEEE. Reprinted, with permission, from [33]

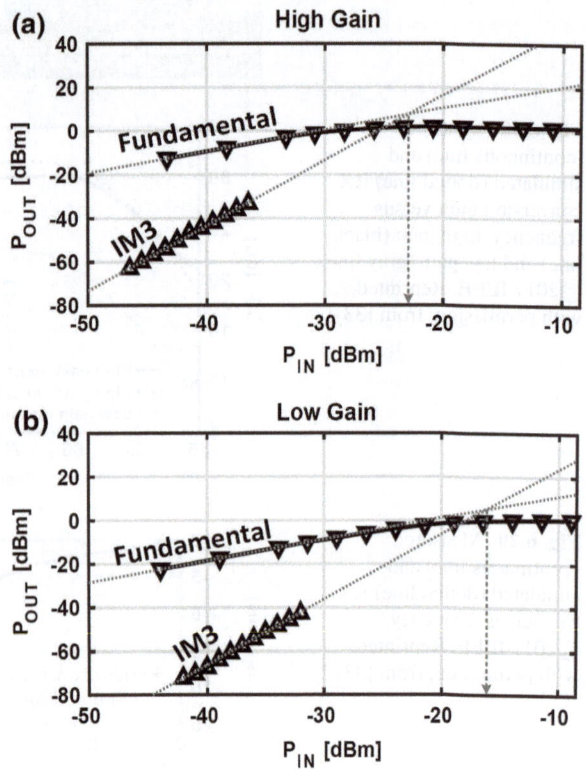

Fig. 6.31 Measured RX IIP3 (1 GHz offset) against input power at 84 GHz, high gain (**a**) and low gain (**b**). ©2017 IEEE. Reprinted, with permission, from [33]

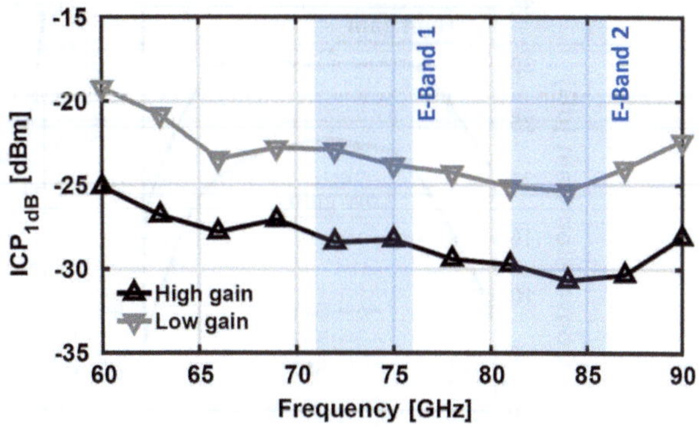

Fig. 6.32 Measured RX input-referred 1 dB compression point versus frequency, high gain (black line) and low gain (gray line). ©2017 IEEE. Reprinted, with permission, from [33]

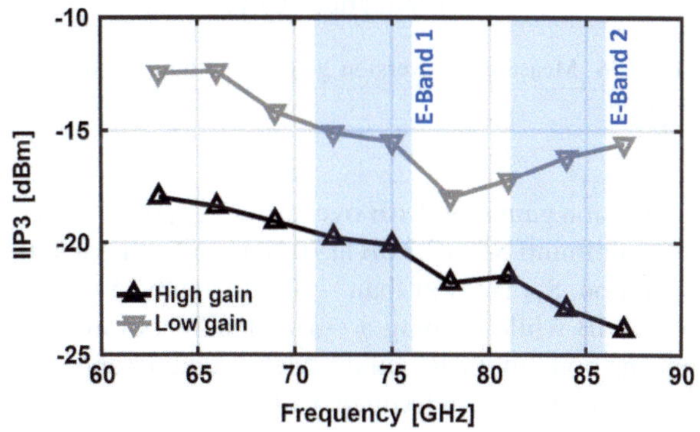

Fig. 6.33 Measured RX IIP3 (1 GHz offset) versus frequency, high gain (black line) and low gain (gray line). ©2017 IEEE. Reprinted, with permission, from [33]

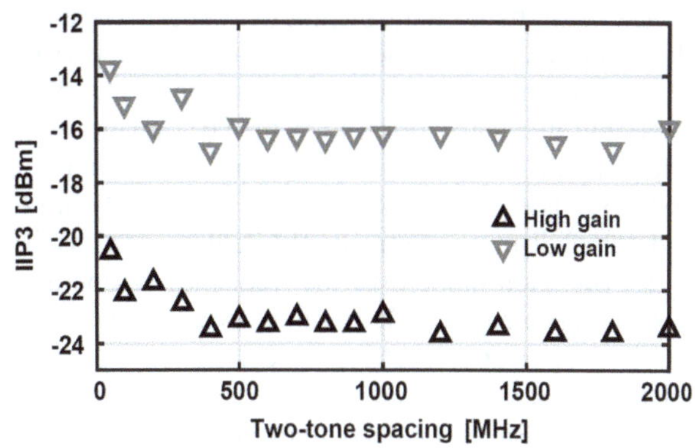

Fig. 6.34 Measured RX IIP3 versus tone spacing at 84 GHz, high gain (black line) and low gain (gray line). ©2017 IEEE. Reprinted, with permission, from [33]

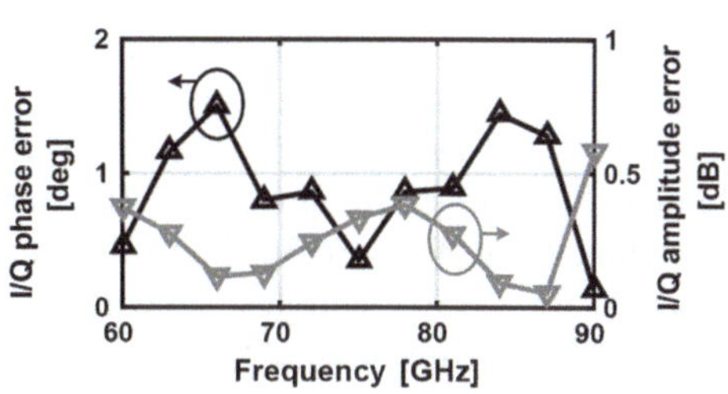

Fig. 6.35 Measured phase and amplitude mismatch between I and Q paths versus frequency

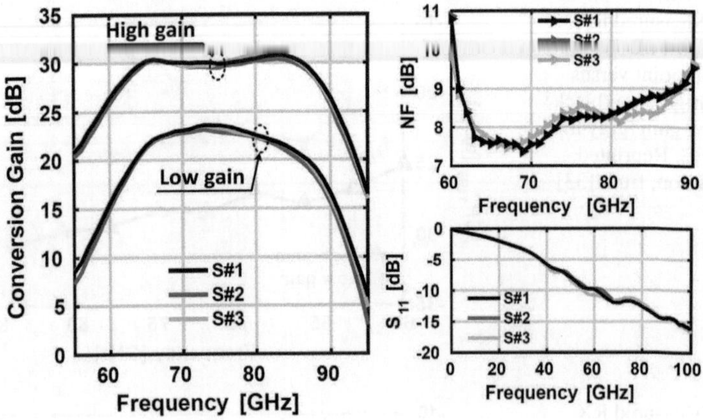

Fig. 6.36 Measured conversion gain, noise figure and input match against frequency for three samples

conversion gain is 30.8 dB over a $BW_{-3\,\mathrm{dB}}$ of 27.5 GHz. In-band gain ripple is <1 dB. The minimum NF is 7.3 dB at 70 GHz and varies less than 2 dB over the whole band of operation. S_{11} is lower than −10 dB from 60.5 to 100 GHz. The gain can be reduced by 7.2 dB while keeping a −3 dB bandwidth in excess of 20 GHz by acting on the biasing point of the RF mixer. Measurements and simulations match very well, especially when the high frequency of operation in combination with the wide band-pass bandwidth is considered.

The IIP3, measured by applying two interferers at 84 GHz with 1 GHz offset is shown in Fig. 6.31. These measurements were performed using a Millitech CMT-12 Magic Tee with two E-Band sources at the input, a 50 Ω termination on one output and a variable attenuator followed by the DUT on the second output. The measured input-referred 1 dB compression point and IIP3 over frequency are shown in Figs. 6.32 and 6.33. The worst cases $ICP_{1\,\mathrm{dB}}$ in the high and low gain mode are −30.7 dBm and −25.3 dBm, respectively. While the worst case IIP3 is 23.8 dBm (18.1 dBm) in high (low) gain mode over the frequency of operation. Two-tone measurements at 84 GHz repeated for different tone spacing (Δf) are reported in Fig. 6.34. The IIP3 is −20.6 dBm at 50 MHz Δf, degrades to −23.1 dBm at 500 MHz Δf and stays relatively flat up to 2 GHz Δf. It is worth noting that a large tone spacing ensures a worst case IIP3, particularly relevant when signal with large modulation bandwidth are used.

I/Q phase and amplitude imbalance, measured by applying the downconverted signal to a sampling oscilloscope, are better than 1.6° and 0.7 dB respectively from 60 to 90 GHz (see Fig. 6.35). The −3 dB output bandwidth at baseband is 1.9 GHz, limited by the off-chip connections. The measured image rejection is better 80 dB over the whole band. Figure 6.36 shows the small signal and noise figure measurements repeated for several samples, showing good repeatability. The block diagrams of measurement setups used are shown in Fig. 6.37.

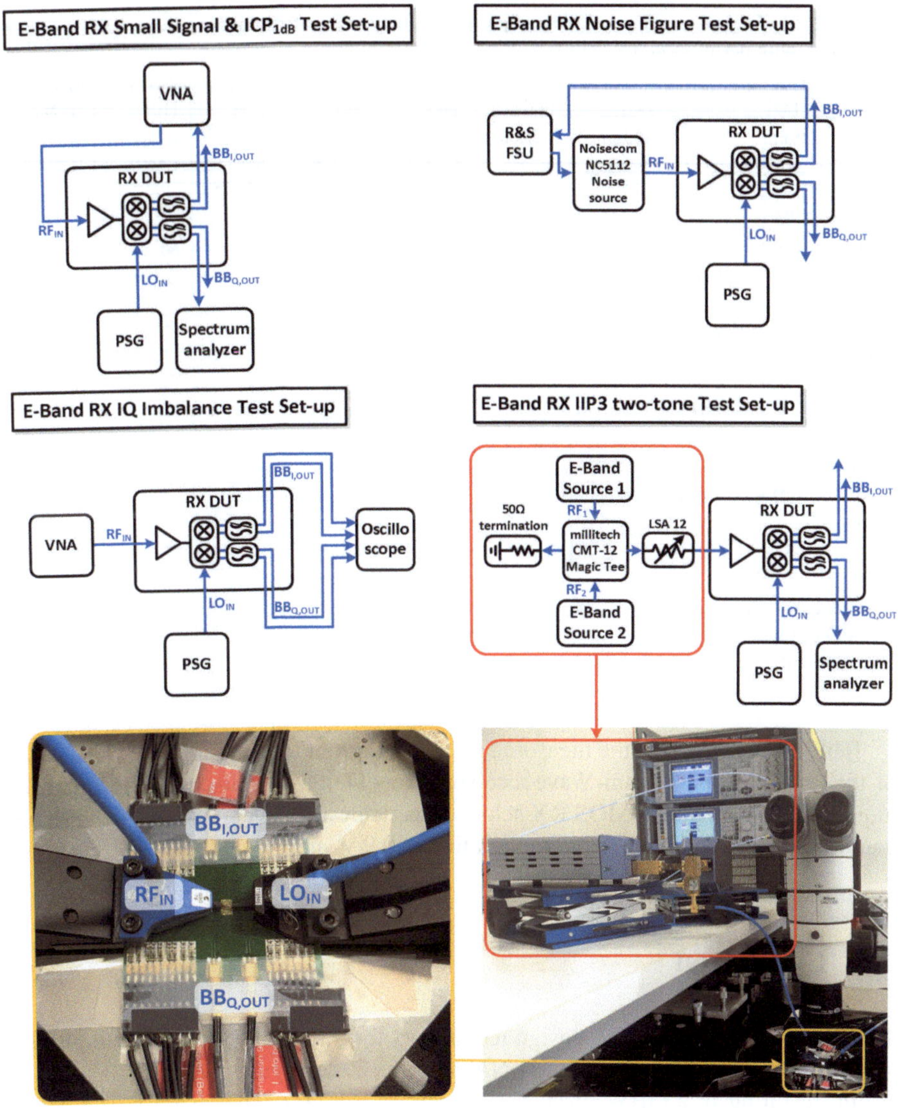

Fig. 6.37 Simplified block diagram of the test set-up used to measure S_{11}, conversion gain, image rejection, $ICP_{1\,\mathrm{dB}}$ (top left), noise figure (top right), I/Q imbalance (center left), IIP3 two-tone test (center right). The mm-Wave RF and LO signals are provided through GSG probes, while the BB I/Q signals and DC pads are wire-bonded to a PCB (bottom)

Table 6.2 Performance summary and comparison with the state-of-the-art receivers. ©2017 IEEE. Reprinted, with permission, from [33]

Ref.	This work [33]		TMTT16 [45]	TMTT16 [45]	ISSCC15 [46]	JSSC16 [49]	JSSC15 [11]	TMTT15 [10]	JSSC11 [17]	
Tech.	**28 nm CMOS**		130 nm Si-Ge	130 nm Si-Ge	28 nm CMOS	40 nm CMOS	45 nm SOI-CMOS	65 nm CMOS	65 nm CMOS	
V_{DD} (V)	**0.9**		1.6–2.7	1.6–2.7	0.9	1.1	1.1	1	1.2	
f_c (GHz)	**75**		73	83	79	61	55	60	60	
Gain (dB)	**30.8**	**23.6**	70	70	35	20	26.2	36	35.5	14
RF-BW (dB)	**27.5**	**21.7**	5	5	8	20	21	7.5	13	13
NF (dB)	**7.3–9.1**	**9.5–12.9**	6–7	6–7	6.2–7	>5.5	5.5–10+	3.8–7	5.6–6.5	n.a.
ICP_{1dB} (dBm)	**−30.7**	**−25.3**	n.a.	n.a.	−32.5	−24	−27	−18	−39	−21
IIP3 (dBm)	**−23.8**	**−18.1**	−10	−12	n.a.	n.a.	n.a.	n.a.	n.a.	n.a.
P_{DC} (mW)	**57**		222	222	59	82.5♯	30*	25	40	

+Graphically estimated
♯No I/Q outputs
*Per element

Table 6.2 summarizes the measured results and provides a comparison with state-of-the-art silicon-based mm-Wave receivers. Benefited by the discussed design techniques, this 28 nm bulk CMOS RX achieves the widest $BW_{-3\,dB}$ with <1 dB in-band gain ripple and <2 dB noise figure variation at 0.9 V supply.

6.6 Conclusion

This chapter has focused on design techniques for mm-Wave low-noise amplifiers and downconverter. The presented LNA achieves 29.6 dB gain over 28.3 GHz bandwidth, resulting in a GBW product in excess of 0.8 THz. The 28 nm bulk CMOS E-Band sliding-IF receiver shows $BW_{-3\,dB}$ from 61.3 to 88.8 GHz with less than 1 dB in-band ripple and <2 dB noise figure variation at 0.9 V supply. This work advances the state-of-the-art and demonstrates the first broadband receiver suitable for E-Band point-to-point communication links in deep-scaled bulk CMOS.

References

1. C. Andrews, A.C. Molnar, A passive mixer-first receiver with digitally controlled and widely tunable RF interface. IEEE J. Solid-State Circuits **45**(12), 2696–2708 (2010)
2. H. Darabi, A. Mirzaei, M. Mikhemar, Highly integrated and tunable RF front ends for reconfigurable multiband transceivers: a tutorial. IEEE Trans. Circuits Syst. I Regul. Pap. **58**(9), 2038–2050 (2011)
3. A. Liscidini, Fundamentals of modern RF wireless receivers: a short tutorial. IEEE Solid-State Circuits Mag. **7**(2), 39–48 (2015)
4. S. Shakib, H.C. Park, J. Dunworth, V. Aparin, K. Entesari, A highly efficient and linear power amplifier for 28-GHz 5G phased array radios in 28-nm CMOS. IEEE J. Solid-State Circuits **51**(12), 3020–3036 (2016)
5. A. Moroni, D. Manstretta, A broadband millimeter-wave passive CMOS down-converter, in *2012 IEEE Radio Frequency Integrated Circuits Symposium*, Montreal, QC (2012), pp. 507–510
6. I. Fabiano, M. Sosio, A. Liscidini, R. Castello, SAW-less analog front-end receivers for TDD and FDD. IEEE J. Solid-State Circuits **48**(12), 3067–3079 (2013)
7. B. Razavi, Design considerations for direct-conversion receivers. IEEE Trans. Circuits Syst. II: Analog Digit. Signal Process. **44**(6), 428–435 (1997)
8. S. Shahramian, Y. Baeyens, N. Kaneda, Y.K. Chen, A 70100 GHz direct-conversion transmitter and receiver phased array chipset demonstrating 10 Gb/s wireless link. IEEE J. Solid-State Circuits **48**(5), 1113–1125 (2013)
9. K. Okada et al., Full four-channel 6.3-Gb/s 60-GHz CMOS transceiver with low-power analog and digital baseband circuitry. IEEE J. Solid-State Circuits **48**(1), 46–65 (2013)
10. H. Wu, N.Y. Wang, Y. Du, M.C.F. Chang, A blocker-tolerant current mode 60-GHz receiver with 7.5-GHz bandwidth and 3.8-dB minimum NF in 65-nm CMOS. IEEE Trans. Microw. Theory Tech. **63**(3), 1053–1062 (2015)
11. S. Kundu, J. Paramesh, A compact, supply-voltage scalable 4566 GHz baseband-combining CMOS phased-array receiver. IEEE J. Solid-State Circuits **50**(2), 527–542 (2015)
12. Behzad Razavi, *RF Microelectronics*, 2nd edn. (Prentice Hall, New Jersey, 2011)
13. B. Razavi, A study of injection locking and pulling in oscillators. IEEE J. Solid-State Circuits **39**(9), 1415–1424 (2004)
14. A. Mirzaei, M. Mikhemar, H. Darabi, 21.8 A pulling mitigation technique for direct-conversion transmitters, in *2014 IEEE International Solid-State Circuits Conference Digest of Technical Papers (ISSCC)*, San Francisco, CA (2014), pp. 374–375
15. D. Zhao, P. Reynaert, A 40 nm CMOS E-band transmitter with compact and symmetrical layout floor-plans. IEEE J. Solid-State Circuits **50**(11), 2560–2571 (2015)
16. A. Mirzaei, H. Darabi, D. Murphy, A low-power process-scalable super-heterodyne receiver with integrated high- Q filters. IEEE J. Solid-State Circuits **46**(12), 2920–2932 (2011)
17. F. Vecchi et al., A wideband receiver for multi-Gbit/s communications in 65 nm CMOS. IEEE J. Solid-State Circuits **46**(3), 551–561 (2011)
18. H.T. Friis, Noise figures of radio receivers. Proc. IRE **32**(7), 419–422 (1944)
19. Low-noise amplifier design techniques, in *IEEE RFIC Virtual Journal* (2014)
20. P. Reynaert, W. Steyaert, M. Vigilante, "RF CMOS". Nanoelectronics: Materials, Devices, Applications, 2 Volumes (2017)
21. T.H. Lee, *The Design of CMOS Radio-Frequency Integrated Circuits* (Cambridge university press, Cambridge, 2003)
22. D. Zhao, P. Reynaert, A 60-GHz dual-mode class AB power amplifier in 40-nm CMOS. IEEE J. Solid-State Circuits **48**(10), 2323–2337 (2013)
23. M. Bassi, J. Zhao, A. Bevilacqua, A. Ghilioni, A. Mazzanti, F. Svelto, A 40–67 GHz power amplifier with 13 dBm P_{SAT} and 16% PAE in 28 nm CMOS LP. IEEE J. Solid-State Circuits **50**(7), 1618–1628 (2015)

24. J. Zhao, M. Bassi, A. Mazzanti, F. Svelto, A 15 GHz-bandwidth 20dBm PSAT power amplifier with 22% PAE in 65nm CMOS, in *2015 IEEE Custom Integrated Circuits Conference (CICC)*, San Jose, CA (2015), pp. 1–4
25. W. Zhuo et al., A capacitor cross-coupled common-gate low-noise amplifier. IEEE Trans. Circuits Syst. II Express Briefs **52**(12), 875–879 (2005)
26. X. Li, S. Shekhar, D.J. Allstot, Gm-boosted common-gate LNA and differential colpitts VCO/QVCO in 0.18μm CMOS. IEEE J. Solid-State Circuits **40**(12), 2609–2619 (2005)
27. W.M.C. Sansen, *Analog Design Essentials*, vol. 859 (Springer Science & Business Media, Berlin, 2007)
28. D. Zhao, P. Reynaert, A 40-nm CMOS E-band 4-way power amplifier with neutralized bootstrapped cascode amplifier and optimum passive circuits. IEEE Trans. Microw. Theory Tech. **63**(12), 4083–4089 (2015)
29. S.Y. Yue, D.K. Ma, J.R. Long, A 17.1-17.3-GHz image-reject downconverter with phase-tunable LO using 3x subharmonic injection locking. IEEE J. Solid-State Circuits **39**(12), 2321–2332 (2004)
30. M. Khanpour, K.W. Tang, P. Garcia, S.P. Voinigescu, A wideband W-band receiver front-end in 65-nm CMOS. IEEE J. Solid-State Circuits **43**(8), 1717–1730 (2008)
31. E. Laskin et al., Nanoscale CMOS transceiver design in the 90170-GHz range. IEEE Trans. Microw. Theory Tech. **57**(12), 3477–3490 (2009)
32. W.L. Chan, J.R. Long, A 5865 GHz neutralized CMOS power amplifier with PAE above 10% at 1-V supply. IEEE J. Solid-State Circuits **45**(3), 554–564 (2010)
33. M. Vigilante, P. Reynaert, On the design of wideband transformer-based fourth order matching networks for e-band receivers in 28-nm CMOS. IEEE J. Solid-State Circuits **52**(8), 2071–2082 (2017)
34. M. Parvizi, K. Allidina, M.N. El-Gamal, A sub-mW, ultra-low-voltage, wideband low-noise amplifier design technique. IEEE Trans. Very Large Scale Integr. (VLSI) Syst. **23**(6), 1111–1122 (2015)
35. W. Sansen, 1.3 analog CMOS from 5 micrometer to 5 nanometer, in *2015 IEEE International Solid-State Circuits Conference - (ISSCC) Digest of Technical Papers*, San Francisco, CA (2015), pp. 1–6
36. H. Wang, C. Sideris, A. Hajimiri, A CMOS broadband power amplifier with a transformer-based high-order output matching network. IEEE J. Solid-State Circuits **45**(12), 2709–2722 (2010)
37. D. Zhao, P. Reynaert, An E-band power amplifier with broadband parallel-series power combiner in 40-nm CMOS. IEEE Trans. Microw. Theory Tech. **63**(2), 683–690 (2015)
38. J. Borremans, P. Wambacq, C. Soens, Y. Rolain, M. Kuijk, Low-area active-feedback low-noise amplifier design in scaled digital CMOS. IEEE J. Solid-State Circuits **43**(11), 2422–2433 (2008)
39. A. Medra, V. Giannini, D. Guermandi, P. Wambacq, A 79GHz variable gain low-noise amplifier and power amplifier in 28nm CMOS operating up to 125° C. Proc. ESSCIRC **2014**, 183–186 (2014)
40. Y.A. Li, M.H. Hung, S.J. Huang, J. Lee, A fully integrated 77GHz FMCW radar system in 65nm CMOS, in *2010 IEEE International Solid-State Circuits Conference - (ISSCC) Digest of Technical Papers*, San Francisco, CA (2010), pp. 216–217
41. A. Mineyama, Y. Kawano, M. Sato, T. Suzuki, N. Hara, K. Joshin, A millimeter-wave CMOS low noise amplifier using transformer neutralization techniques. Asia-Pac. Microw. Conf. **2011**, 223–226 (2011)
42. S. Onoe, 1.3 Evolution of 5G mobile technology toward 1 2020 and beyond, in *2016 IEEE International Solid-State Circuits Conference (ISSCC)*, San Francisco, CA (2016), pp. 23–28
43. J. Wells, Multigigabit Microwave and Millimeter-Wave Wireless Communications. Artech House (2010)
44. Z. Huang, H.C. Luong, B. Chi, Z. Wang, H. Jia, 25.6 A 70.5-to-85.5GHz 65nm phase-locked loop with passive scaling of loop filter, in *2015 IEEE International Solid-State Circuits Conference - (ISSCC) Digest of Technical Papers*, San Francisco, CA (2015), pp. 1–3

45. R. Levinger, R.B. Yishay, O. Katz, B. Sheinman, N. Mazor, R. Carmon, D. Elad, High-performance E-band transceiver chipset for point-to-point communication in SiGe BiCMOS technology. IEEE Trans. Microw. Theory Tech. **64**(4), 1078–1087 (2016)
46. D. Guermandi, Q. Shi, A. Medra, T. Murata, W. Van Thillo, A. Bourdoux, P. Wambacq, V. Giannini, 19.7 A 79GHz binary phase-modulated continuous-wave radar transceiver with TX-to-RX spillover cancellation in 28nm CMOS, in *2015 IEEE International Solid-State Circuits Conference - (ISSCC) Digest of Technical Papers*, San Francisco, CA (2015), pp. 1–3
47. C.H. Li, C.N. Kuo, M.C. Kuo, A 1.2-V 5.2-mW 20–30-GHz wideband receiver front-end in 0.18-μm CMOS. IEEE Trans. Microw. Theory Tech. **60**(11), 2709–2722 (2011)
48. M. Vigilante, P. Reynaert, A 25-102GHz 2.81-5.64mW tunable divide-by-4 in 28nm CMOS, in *2015 IEEE Asian Solid-State Circuits Conference (A-SSCC)* (2015), pp. 1-4
49. V. Bhagavatula, T. Zhang, A.R. Suvarna, J.C. Rudell, An ultra-wideband IF millimeter-wave receiver with a 20 GHz channel bandwidth using gain-equalized transformers. IEEE J. Solid-State Circuits **51**(2), 323–331 (2016)

Chapter 7
mm-Wave Highly-Linear Broadband Power Amplifiers

In Chap. 1 the system level requirements in terms of TX bandwidth and linearity were addressed. Chapter 2 introduced the design challenges of deep-scaled CMOS actives and passives components and in Chap. 3 several gain-bandwidth enhancement techniques were considered. This chapter brings these results together and applies them to mm-Wave broadband CMOS power amplifiers (PAs) design.

The basics of PA design and the challenges specific to mm-Wave operation are recalled in Sect. 7.1. Some of the intricacies of class-AB operation such as efficiency at power back-off, major causes of AM-PM distortion and linearization techniques are the focus of Sect. 7.2. Finally, Sect. 7.3 discusses the design, layout and measurement details of a 29–57 GHz (65% BW) AM-PM compensated class-AB power amplifier tailored for 5G phased arrays. Designed in 0.9 V 28 nm CMOS without RF thick top metal, the PA achieves a P_{sat}=15.1 dBm $\pm$ 1.6 dB and |AM-PM| $<$ 1° from 29 to 57 GHz, with a peak PAE of 24.2%. Techniques are studied to realize the required load impedance and distortion cancellation over the wide band of operation, while allowing 2-way power combining to further increase the delivered P_{OUT}. The very low AM-PM distortion of the realized PA enables up to 10.1, 8.9, 5.9 dBm average P_{OUT} while amplifying a 1.5, 3, 6 Gb/s 64-QAM respectively at 34 GHz with EVM/ACPR better than −25 dB/−30 dBc, without any digital pre-distortion.

The results of the mm-Wave highly-linear broadband power amplifier in Sect. 7.3 have been published in IEEE Radio Frequency Integrated Circuits Symposium (RFIC 2017) and in the IEEE Journal of Solid-State Circuits (JSSC 2018, vol. 53, no. 5).

M. Vigilante and P. Reynaert, *5G and E-Band Communication Circuits in Deep-Scaled CMOS*, Analog Circuits and Signal Processing,
https://doi.org/10.1007/978-3-319-72646-5_7

7.1 Power Amplifiers Basics

7.1.1 Single Transistor Amplifier Under Large Signal

Figure 7.1 shows the simulated output current of a 42 × 12 μm/28 nm common source amplifier terminated on a short circuit versus input/output DC voltage. Clearly I_O is a non-linear function of both the input and output voltage. In turns, the output voltage depends on the output current and the impedance seen at the output node at all the non-negligible current harmonics. Moreover, since a power amplifier is an amplifier, we expect that when the input signal is large, the output signal is even larger.

Therefore, PA design is a problem that does not lead to trivial solutions. And more than often PA designers rely heavily on extensive simulations and complex measurement setups, such as load-pull and modulated signal. If on the one hand these tools are necessary, on the other it is important to start from the basics to get insight into the fundamental problems and limitations.

7.1.2 Trade-Offs in PA Design: P_o, PAE and Linearity

The power amplifier should deliver sufficient power to the typically 50 Ω antenna, while ensuring sufficient in-band and out-of-band linearity (i.e. EVM and ACPR) to fulfill the system level requirements discussed in Chap. 1.

To get insight into the basic trade-offs of PA design, it is instructive to refer to the simplified schematic in Fig. 7.2. Assuming that the input and output matching networks resonate at the frequency of operation, the power delivered to the load R_L is simply

$$P_L = V_{L,rms} I_{L,rms} = \frac{V_{L,rms}^2}{R_L} = \frac{(n\, V_{o,rms})^2}{R_L}. \tag{7.1}$$

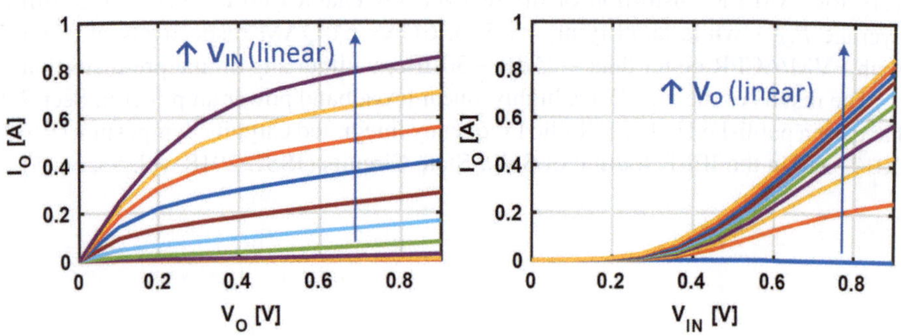

Fig. 7.1 Simulated output current of a 42 × 12 μm/28 nm common source amplifier terminated on a short circuit versus input/output DC voltage

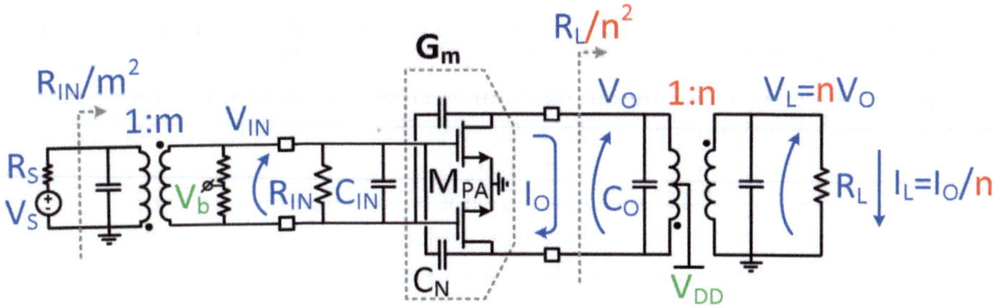

Fig. 7.2 Simplified schematic of a pseudo-differential neutralized CS power amplifier with transformer-based matching networks

Transformers are instrumental to resonate the parasitic capacitance of the PA, while providing a center tap for DC biasing, differential to single ended transformation, galvanic isolation and impedance transformation. The latter property is particularly relevant in deep-scaled CMOS.

To ensure reliable operation and prevent breakdown, the supply voltage scales together with the minimum feature size (see Chap. 1). For any PA topology, the maximum linear output voltage (V_O in Fig. 7.2) is always limited by and proportional to V_{DD}. Therefore, to deliver the required output power the current needs to increase. This is why an impedance transformation network is normally required at the output of a PA, realized with a 1:n ideal transformer in Fig. 7.2. Clearly, the transistor size needs to be increased accordingly to deliver the required current.

Any PA needs DC power to deliver power at RF frequency. How efficiently this conversion happens is measured by power added efficiency (PAE) defined as

$$PAE = \frac{P_L - P_S}{P_{DC}}. \tag{7.2}$$

Any power dissipated by the transistor and by the lossy matching networks results in a penalty in terms of PAE.

The first way to enhance the PAE in Eq. 7.2 is obviously to reduce the DC power consumption by working on the biasing point (V_b in Fig. 7.2). This is indeed an extremely effective solution, at least to some extend. Clearly if $V_{DD} = 0V$, $P_{DC} = 0W$ the PA is off and no power is even delivered. In a classical textbook several PA classes are defined for linear PAs depending on the biasing point (e.g. class-A, class-AB, class-B and class-C) [1, 2]. A few remarks to be addressed at this point are the following. In a class-A power amplifier the output signal is not allow to compress, posing a limitation to the maximum input voltage that can be applied, and therefore creating doubts on the practical existence of this class of PAs in the first place. A PA biased in class-C should be biased below threshold at 0 A bias current. In deep-scaled CMOS the threshold voltage is vaguely defined and the sub-threshold current of an ultra-low-V_t device normally used for mm-Wave applications may be not negligible at all. Therefore, practical state-of-the-art PAs are normally biased in class-AB. The real question is how deep in class-AB should be biased the PA for best performance.

Another way to improve the PA efficiency is to reduce the power dissipated in the transistor. As it will be shown shortly, this can be achieved by accurately design the output load impedance at different harmonics. However, these powerful techniques do have some shortcomings. (1) To achieve proper voltage shaping the harmonic content of the current should be substantial, meaning that the PA should already be close to compression. (2) Harmonic tuning normally requires a high order tank with high quality factor, which is not always possible to realize in bulk CMOS. And (3) harmonics at the output of any amplifier are synonym of distortion, posing doubts on the effect on linearity.

These considerations lead us to one of the most classical trade-off in PA design, PAE versus linearity. We would like the PA to be biased in deep class-AB, to benefit from substantial DC power saving and, as it will be shown shortly, of gain expansion, further improving PAE and 1dB compression point. On the other hand, we expect that as we move close to the threshold voltage, the device non-linearity will be exacerbated, generating more distortion and harmonics. So, distortion is good for efficiency, and perhaps not surprisingly, bad for linearity.

It is interesting to highlight at this point that different distortion mechanisms may also result in distortion cancellation, yielding superior efficiency for the same or even better linearity at high power levels [1–4]. However, those fortuitous biasing points may depends heavily on PVT variations resulting in excellent performance in a well controlled lab environment but more difficult to exploit in a real application [5].

7.1.3 Harmonic Terminations and Switching Amplifiers

A power amplifier is an amplifier designed to work under large signal regime. Therefore, the harmonic content of the output current will be rich even if the input is a clear single tone. As mentioned earlier, the power dissipated in the transconductor results in a penalty in PAE. We therefore seek for a load Z_L that could exploit the current harmonics to properly shape the output voltage and minimize this kind of power loss. To get insight it is useful to refer to the simplified schematic shown in Fig. 7.3. Here, the transistor is biased in ideal class-B, $I_{DC} = 0A$ and V_{IN} is an ideal sine. When $V_{IN} > V_{th}$ the transistor turns on in saturation and $I_O = G_m V_{IN}$. When $V_{IN} \leq V_{th}$, $I_O = 0A$. We further assume that the transconductor never enters the triode region, therefore its output current depends on the input voltage only. This is clearly far from reality, nevertheless it is useful to simplify the problem and get a basic understanding of the circuit operation.

Figure 7.3 shows also the transistor output current and voltage waveforms in the time domain when the load provides high impedance at the odd harmonics. Clearly, if I_O and V_O do not overlap, no power is dissipated in the active device and a theoretical 100% efficiency could be achieve. Any good paper about power amplifiers starts with a theoretical discussion on PAE, a comparison with class-B amplifier is provided and 100% efficiency is claimed. As it will be shown shortly, there are several reasons why such good performance has never been measured, and designs at mm-Wave do not get even close to 50% efficiency.

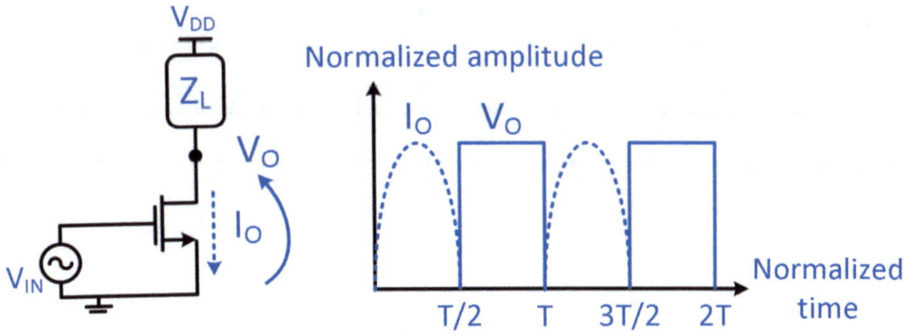

Fig. 7.3 Simplified schematic of a CS power amplifier with highlighted device output current and voltage

Figure 7.4 shows the ideal current and voltage waveforms at the output of the active device when the load is designed to provide specific impedance at the harmonics. The output current can be written as [6]

$$I_O = |I_{\omega_o}| + |I_{2\omega_o}|e^{j90^\circ} + |I_{3\omega_o}| \tag{7.3}$$

where the odd harmonics are in-phase with the input voltage and the even harmonics are 90° out of phase. Harmonics higher than the 3rd are neglected for simplicity. The output voltage depends on the output current and the load impedance, in this simplified example it can be written as

$$\begin{aligned} V_O = &-|I_{\omega_o}||Z_{L\omega_o}|e^{j\angle Z_{L\omega_o}} - |I_{2\omega_o}||Z_{L2\omega_o}|e^{j(90^\circ + \angle Z_{L2\omega_o})} + \\ &- |I_{3\omega_o}||Z_{L3\omega_o}|e^{j\angle Z_{L3\omega_o}}. \end{aligned} \tag{7.4}$$

When the load impedance is tuned at the fundamental frequency and the quality factor is high enough to shunt all the harmonics of the current to ground, ideal class-B operation is achieved and the resulting PA waveforms are shown in Fig. 7.4a. When the matching network is lossless and the knee voltage is neglected, an ideal 78.5% efficiency can be achieved [1]. This is the simpler PA to design and shows a remarkable PAE, several effort have been made over the years to beat this number, the most successful of them are reported in the following.

When a strong 2nd harmonic component 90° out of phase with respect to the fundamental is added to the output voltage, the waveform shows a narrower asymmetric shape as shown in Fig. 7.4b, c. This is a particularly interesting effect, given the fact that the second harmonic component of the current is by far the stronger one in a class-B amplifier [1]. The class-J operation leverages this effect to improve the efficiency of a practical matching network while keeping the maximum theoretical 78.5% efficiency. In this case, the load impedance is allowed to show a capacitive termination at the 2nd harmonic, which can greatly help in a practical case. To keep the necessary 90° relation in Eq. 7.4 the load impedance needs to show an inductive

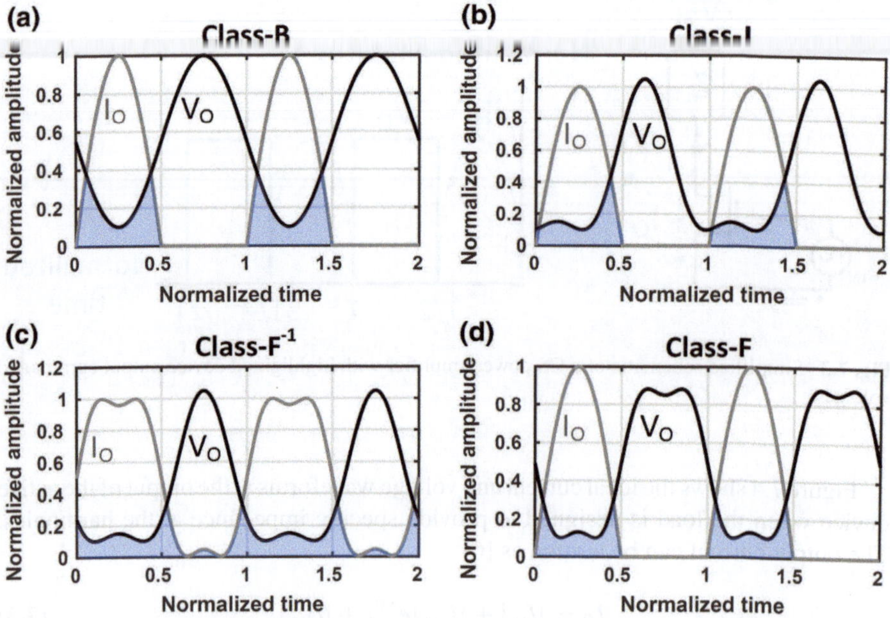

Fig. 7.4 Ideal device output current and voltage for ideal operation in class-B (**a**), class-J (**b**), class-F^{-1} (**c**) and class-F (**d**) with highlighted the power dissipated in the transconductor. A knee voltage of 10% of the fundamental voltage component is imposed

termination at the fundamental frequency. This is the reason why in this amplifier the output current and voltage are not 180° out of phase (see Fig. 7.4b). This is also the reason why this design is prone to a substantial increase of AM-PM distortion when compared to the class-B implementation, as it will be discussed later.

Two circuits that do theoretically better are the class-F^{-1} and the class-F shown in Fig. 7.4c, d. The former requires a high impedance at the 2nd harmonic and a low impedance at the 3rd one. The contrary is required for ideal class-F operation. Both of them can achieve an efficiency theoretically as high as 81.6% when the knee voltage is neglected. Clearly, the class-F^{-1} is very similar to a class-J. However, in this case the output voltage and current are 180° out of phase, with a benefit in efficiency and AM-PM distortion, as it will be shown later. Another interesting difference with class-J is that theoretically the current waveform should be close to a square. It is obvious that such a current shape can appear if and only if the amplifier is very close to compression, possibly deep in compression, switching between on and off state.

It is worth noting that in a differential circuit such as the one shown in Fig. 7.2 the odd harmonic component of the current flow differentially, while the even harmonics flow in common mode. This property can be leveraged during design phase as previously discussed for integrated oscillators in Chap. 4. Moreover, higher order tanks as the one presented in Chap. 3 can be used to realize broadband class-B amplifiers or narrowband class-F amplifiers.

Finally, a class of amplifiers widely adopted in the sub-GHz range are the switching amplifiers, such as class-D and class-E [1]. Provided that good switches are available, these amplifiers achieve measured efficiencies close to 100% while being completely non linear. To restore linearity other techniques need to be applied, such as outphasing or envelope tracking [7]. However, (1) it has been proven both theoretically and with measurements that when the frequency of operation exceeds 2% of f_t while $R_L/R_{ON} < 100$ the benefit of switching power amplifiers vanishes rapidly [8–10]. And (2) when broadband modulation schemes are to be transmitted, as it often is the case for mm-Wave 5G systems, both outphasing and envelope tracking are increasingly difficult to realize. Therefore, this last class of amplifiers is of little use at mm-Wave frequencies.

7.1.4 Challenges @mm-Wave

As discussed in Chap. 1 transistors are getting faster (Fig. 1.4), on the other hand gain is rapidly decreasing at mm-Wave frequencies (Fig. 1.5) while the voltage supply scales with minimum channel length (Fig. 1.6). Moreover, the maximum linear output power of any class of power amplifiers is proportional to and limited by V_{DD}. The combination of these factors results in the state-of-the-art PAE versus frequency of operation shown in Fig. 7.5.

Clearly, PAE is decreasing with frequency. Moreover, the limited supply voltage requires either a larger impedance transformation, or extensive power combining to achieve the same specifications on delivered output power. However, there are practical limitations on the impedance scaling that can be achieved and the numbers of combining paths. The reasons are the following, (1) matching networks with a large scaling factor normally results in large insertion loss. (2) To keep the same output power at lower V_{DD}, the current delivered by the active devices needs to

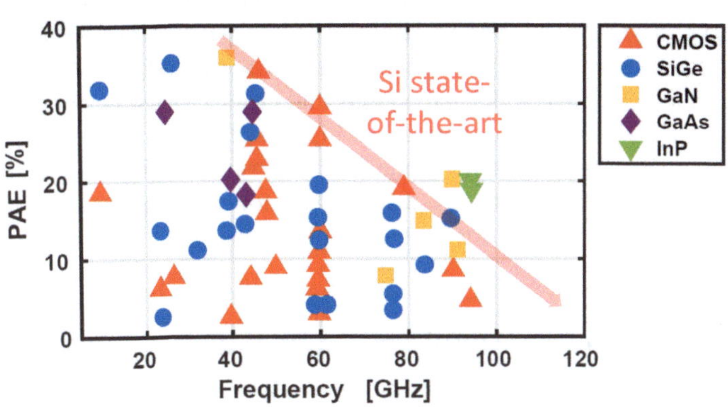

Fig. 7.5 Efficiency of recently published state-of-the-art PAs versus frequency [11]. ©2017 John Wiley and Sons. Reprinted, with permission, from [12]

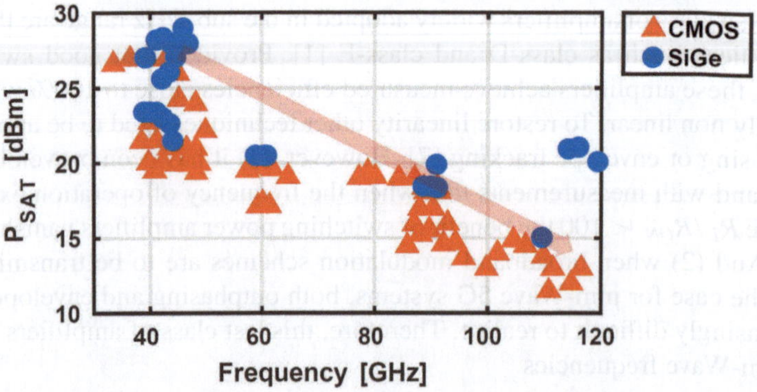

Fig. 7.6 Saturated output power of recently published state-of-the-art PAs versus frequency [14]

increase, therefore the transistor W need to increase too. At mm-Wave the long interconnections of such big devices may easily become dominant sources of power loss [13]. And (3), the number of combining paths increases the insertion loss of the power combiner.

For all these reasons, the delivered output power is decreasing with frequency too, as clear from Fig. 7.6. State-of-the-art solutions normally employ output matching networks that provide a load impedance in the proximity of ≈30 to ≈50 Ω, and two to four combining paths to maximize the output power without compromising PAE [13, 15].

It is worth mentioning that the work in [16] has shown that a PA with ≈10 dBm average output power at −25 dB EVM would allow ≈50 m link distance for a 28 GHz transceiver when a 64-QAM OFDM signal with 250 MHz RF bandwidth and 9.6 dB PAPR is used. Such an EVM allows 3 dB margin on the required SNR. The reference [16] focuses on handset applications and the relatively low output power of the single PA allows to achieve the required link budget thanks to a 16-element beamforming array at both TX and RX side.

To maximize the data rate, 5G wireless links will adopt high order modulation schemes (e.g. 64-QAM) with large RF bandwidth (>100 MHz). At the transmitter side, this implies several design challenges. (1) A wideband PA is needed to cover several channels, amplify wideband signals and ensure robust performance against PVT variations. (2) Modulated signals with high spectral efficiency show large peak-to-average-power-ratio, challenging the linearity versus efficiency trade-off for a given average P_{OUT}. (3) Digital pre-distortion is not easily applicable when several PAs are integrated in an array for handset applications [16].

7.2 Class-AB Power Amplifier @mm-Wave

As clear from the study outlined so far, any power amplifier at mm-Wave frequencies is going to be biased in class-AB. Even a switching amplifier, will not operate close to a switch for most of the time at these high frequencies [8, 10]. Even a class-F or a class-F^{-1} when not operated in deep compression, will behave like a linear class-AB amplifier with a load impedance engineered to show particularly high or low values at the harmonics [9, 11]. Moreover, at mm-Wave wideband modulation schemes are going to be adopted, posing enormous challenges on practical implementation of digital pre-distortion, envelope tracking or outphasing techniques. Like it or not, linear amplifiers (i.e. class-AB) show significant benefits at system level and are so far state-of-the-art PAs at mm-Wave.

7.2.1 *Efficiency at Power Back-Off*

We are still left with a question, how deep in class-AB should we bias the active transistors? To gain insight and intuition on the effect of the bias point in a mm-Wave class-AB PA designed in deep-scaled CMOS, it is useful to introduce the following design example. We start with the simplified class-AB PA schematic in Fig. 7.7a. The input matching network realizes impedance scaling to 50 Ω and provides a center tap to supply the bias voltage. The output matching network is realized with an explicit 30 Ω resistor and an inductor L_L that resonate the PA output capacitance and provides a center tap. The amplifier is tuned at 30 GHz and the transistor are designed with $(W/L)_{PA} = 6 \times 42\,\mu m/28\,nm$.

Let us first consider a classical neutralized common source amplifier (Fig. 7.7b) as transconductance stage. We compare in Fig. 7.8 the effect of two different biasing

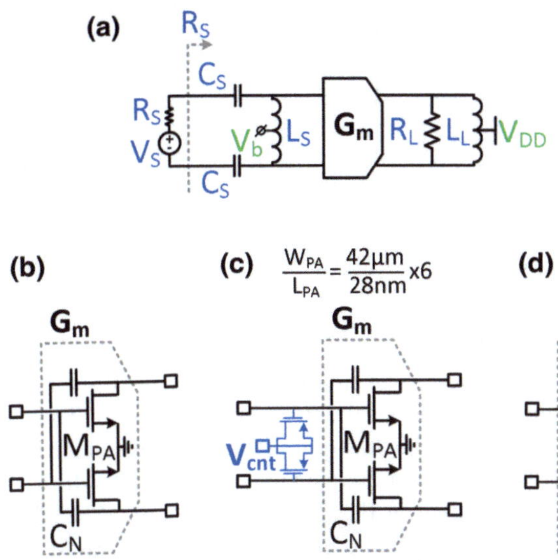

Fig. 7.7 Simplified class-AB power amplifier schematic (**a**). Classical neutralized CS amplifier (**b**), input varactors for non-linear C_{GS} correction (**c**) and degeneration inductance L_{DEG} for linearization (**d**)

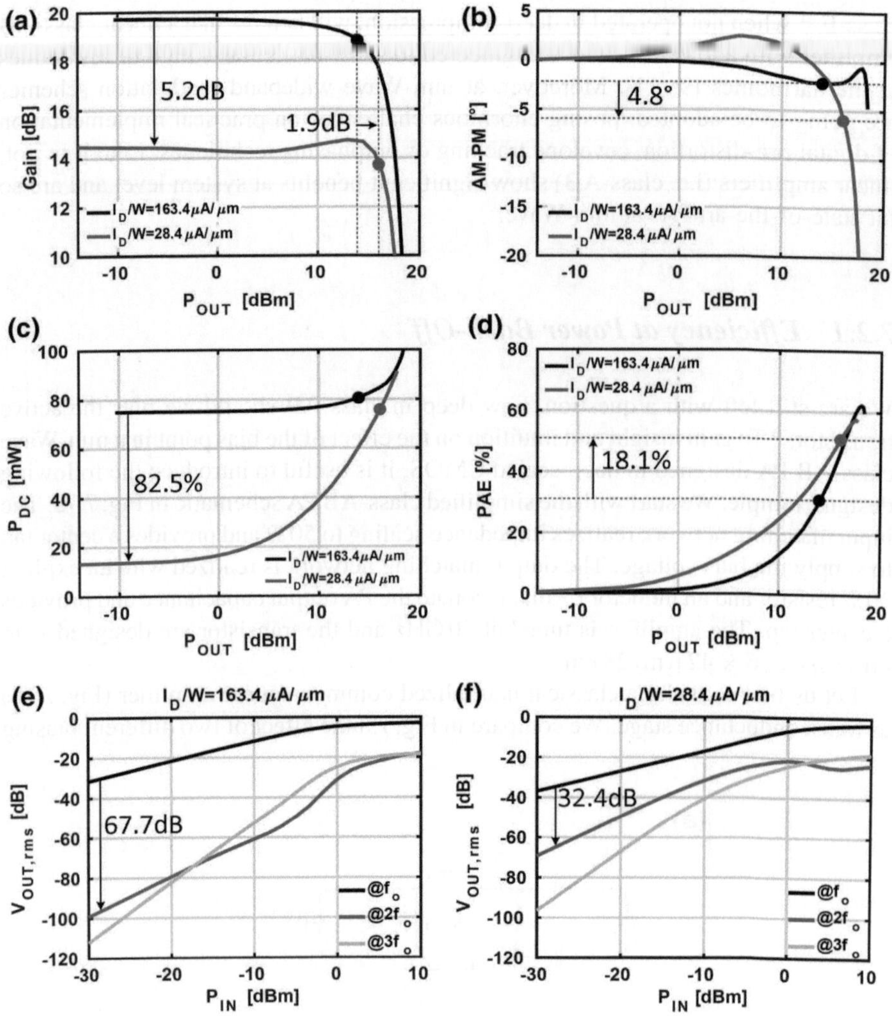

Fig. 7.8 Comparison between different class-AB bias points. **a** gain, **b** AM-PM distortion, **c** P_{DC} and **d** PAE versus P_{OUT}. Harmonic components of the output voltage versus P_{IN} for $I_D/W = 163.4\,\mu A/\mu m$ (**e**) and $I_D/W = 28.4\,\mu A/\mu m$ (**f**)

points on the main PA parameters. The first bias point is $I_D/W = 163.4\,\mu\text{A}/\mu\text{m}$ provides best $f_t\, g_m/I_D$ and it is a classical choice in LNA design. The second bias point is $I_D/W = 28.4\,\mu\text{A}/\mu\text{m}$. Such a low quiescent current results in several interesting effects. First, 5.2 dB gain reduction, gain expansion and 1.9 dB higher output 1 dB compression point are evident in Fig. 7.8a. Therefore the AM-AM distortion is improved at expenses of small signal gain. However, the AM-PM distortion in Fig. 7.8b is 4.8° worse at $P_{1\,dB}$ and also remarkably worse before $P_{1\,dB}$, where the amplifier is suppose to be linear. The DC power consumption in Fig. 7.8c shows a 82.5% improvement at low power levels that remains lower even at $P_{1\,dB}$. This result in the remarkable improvement in PAE shown in Fig. 7.8d, both at $P_{1\,\mathrm{dB}}$ and at back-off. From another prospective, we can notice the much higher harmonic content of the output voltage in Fig. 7.8e, f.

We further notice that in this simplified design example we did not provide any harmonic control at the output. It should be noticed that in the case of a class-J or class-F^{-1} PA, the load impedance at the second harmonic should be high, and provide the proper 90° phase shift (as shown in Eq. 7.4). This is particularly challenging in this case, since the 2nd harmonic flows in common mode. An in CM the PA is not well input-output isolated, and the neutralization capacitor makes everything much worse, as discussed in Chap. 6. Moreover, at the 2nd harmonic capacitors show 2× lower impedance. Therefore, the load impedance at the 2nd harmonic is particularly sensitive to the input load. Finally, even if Z_L realizes an high impedance at the 2nd harmonic, any low impedance in parallel with it (coming from the input load for instance) would compromise the proper class-J or class-F^{-1} operation of the PA.

From the single-tone continuous-wave performance summarized in Fig. 7.8 it is easy to conclude that as long as the amplifier gain does not drop too much, a deep class-AB operation results in much better performance at expenses of more distortion. However, in a real application the PA needs to amplify wideband modulated signals with large PAPR. Therefore, a back-off from $P_{1\,\mathrm{dB}}$ is needed to guarantee the specifications on linearity (both EVM and ACPR). Furthermore, when a modulated signal is amplified, second order effects such as memory effects and modulation on the supply rails come into the picture. Those effects are extremely challenging to simulate and simple models based on AM-AM and AM-PM only fails to give an accurate estimate of the EVM degradation under these conditions [16]. This is not the case in presence of AM-AM only (i.e. if AM-PM is negligible) [1, 2]. In the following, we will briefly recall the major mechanisms that generate AM-PM and the most popular state-of-the-art circuit linearization solutions. As it will be shown, higher linearity results in lower asymmetries in the output spectrum, better EVM and ACPR for given average output power at the expenses of lower efficiency.

7.2.2 *Sources of AM-PM Distortion*

The major causes of AM-PM distortion in class-AB power amplifiers have been highlighted in several works [1, 2, 17, 18]. Here we will briefly summarize them, with a strong focus on the most relevant effects in practical class-AB PAs implemented in deep-scaled CMOS.

To get insight, it is useful to refer to the simplified schematic shown in Fig. 7.7a, where a class-AB PA follows an ideally linear driver, $W_{DR} = W_{PA}/2$. It is interesting to note that even if the driver is linear, the signal V_{IN} is distorted even before amplification due to the non-linear input capacitance of the PA C_{IN}. When C_{GD} is perfectly canceled, $C_{IN} = C_{GS}$. If this is not the case, C_{GD} is reflected at the input due to the miller effect, amplified by a factor $(1 - A_V) = (1 + G_m(R_o//R_L))$. Clearly, even in the most simplified large-signal amplifier model, G_m is strongly non-linear.

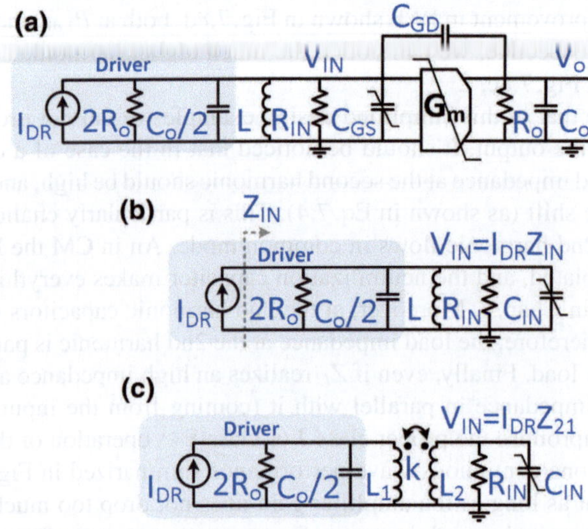

Fig. 7.9 Simplified schematic of a PA with a driver and inter-stage matching network (**a**). Simplified schematic of a narrowband (**b**) and broadband (**c**) inter-stage matching network to study the effect of a non-linear C_{IN} on AM-PM distortion

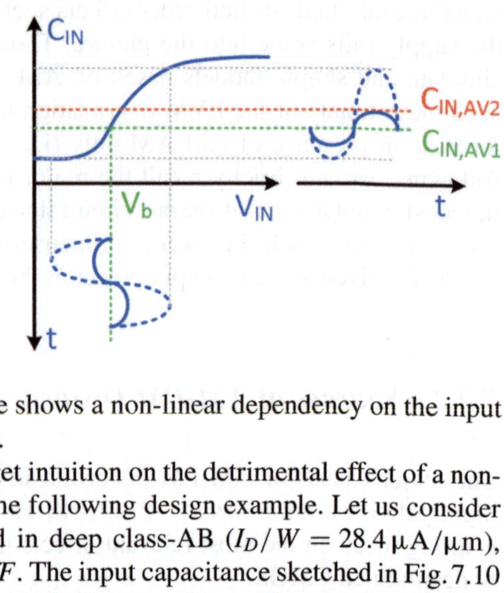

Fig. 7.10 Variation of average input capacitance $C_{IN,AV}$ due to the non-linear asymmetric dependency of C_{IN} on V_{IN} under large-signal operation [17]

Therefore, the miller input capacitance shows a non-linear dependency on the input voltage even if C_{GD} is perfectly linear.

To quantify this phenomenon and get intuition on the detrimental effect of a non-linear C_{IN} on AM-PM we introduce the following design example. Let us consider the active stage in Fig. 7.2b is biased in deep class-AB ($I_D/W = 28.4\,\mu A/\mu m$), $R_{IN} \approx 650\,\Omega$, $R_o \approx 140\,\Omega$, $C_o \approx 116 fF$. The input capacitance sketched in Fig. 7.10 is a non-linear asymmetric function of the input voltage. The simulated values for $C_{IN,min}$ and $C_{IN,MAX}$ after parasitic extraction are 124.5 fF and 153 fF respectively. Under small-signal operation the average input capacitance is equal to the input capacitance at the bias point $C_{IN,AV1} \approx 128 fF$. When a large signal is applied, due to the asymmetries in the voltage to capacitance transfer function, the average value of C_{IN} increases. In this example we set the value of $C_{IN,AV2}$ to $\approx 133 fF$. This is the value of the input capacitance when the PA is biased with

$I_D/W = 163.4\,\mu\text{A}/\mu\text{m}$, a value slightly higher than the 1 dB compression point of this PA as shown in Fig. 7.8c. The variation of the input capacitance with the input voltage results in phase and amplitude distortion (i.e. AM-PM and AM-AM), the latter being negligible. In [16, 18] it is proven that narrowband inter-stage matching networks are exacerbating this mechanism of AM-PM distortion. To investigate how critical is this effect for a mm-Wave 28 nm CMOS PA we focus on the two simplified schematics depicted in Fig. 7.9b, c. In a narrow band inter-stage matching network (Fig. 7.9b) L resonates with $C_o/2 + C_{IN,AV1}$ at the center frequency $f_o = 30\,\text{GHz}$. The broadband matching network (Fig. 7.9c) is designed with $L_1 = 1/((2\pi f_o)^2(C_o/2)(1-k^2))$ and $L_2 = 1/((2\pi f_o)^2 C_{IN,AV1}(1-k^2))$, as discussed in Chap. 3. The resulting input voltage is $V_{IN} = I_{DR}Z_{IN}$ in the former case and $V_{IN} = I_{DR}Z_{21}$ in the latter. When the input voltage amplitude is large, the two circuits will be mistuned due to the increased input capacitance $C_{IN,AV2}$. The effect on AM-PM is shown in Fig. 7.11. Clearly, a broadband inter-stage matching

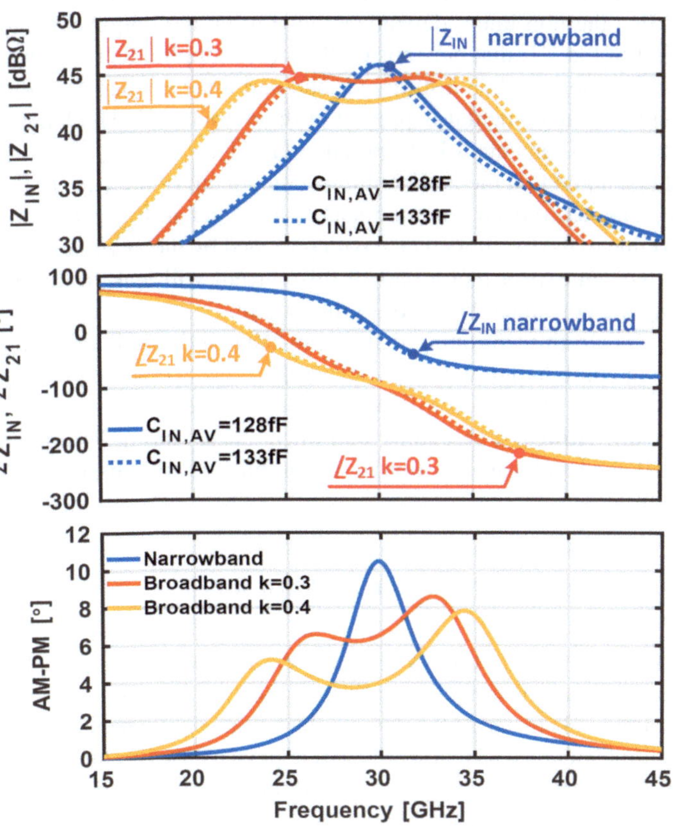

Fig. 7.11 AM-PM distortion due to a variation of $C_{IN,AV}$ in a narrowband versus broadband inter-stage matching network. The bottom graph is directly derived from the middle one by subtracting the phase of Z_{IN} and Z_{21} for the two values of $C_{IN,AV}$ considered. ©2018 IEEE. Reprinted, with permission, from [19]

network shows a remarkable improvement in AM-PM distortion over the whole bandwidth $BW_{-3\,dB}$.

Furthermore, since the non-linear input capacitance distorts the signal prior amplification, the PA can be seen as a cascade of two non linear elements (i.e. C_{IN} and G_m). Therefore, in presence of a modulated signal, the total 3rd order intermodulation distortion will be generated (1) by the non-linear C_{IN} and amplified by G_m, but also (2) by the second order modulation distortion due to C_{IN} amplified by the non-linear G_m.

All these reasoning's point to one preferred direction, de-Qing the inter-stage matching network, adopting a linear driver and possibly compensate for the non-linearity of C_{IN} to minimize AM-PM distortion. Clearly, this will not help efficiency.

Another major cause of AM-PM distortion is, obviously, the non-linear G_m. Since the output current depends non linearly both on the input and the output voltage (as shown in Fig. 7.1), this contribution is the most difficult to analyze. In particular, when the 1st harmonic component of the voltage is out of phase with the fundamental (i.e. the load impedance is not perfectly resistive at the fundamental frequency, as for the class-J PA) the non-linear dependency of G_m on the output voltage will result in a variation of the $\angle I_o$ that depends on the signal amplitude, resulting in AM-PM [18]. The same applies when a strong second harmonic voltage component 90° out of phase with the fundamental is present at the output [18]. The latter being a key condition for class-J and class-F^{-1} operation [1, 11].

It should be noted that the 2nd harmonic component of the current flows in common mode. The neutralized common source amplifier in Fig. 7.7b is not well isolated in CM due to the low impedance path provided by the $2C_{GD} + 2C_N$ capacitors. Therefore, we can expect that a part of this current will leak to the input, further exacerbating the non-linear behavior of this circuit.

Finally, it is worth mentioning that at mm-Wave a driver is normally needed, if not a driver in combination with a pre-driver [13, 20]. These amplifiers would also substantially benefit from a non-linear operation to save DC power, especially at power back-off. However, cascading non-linear stages has a detrimental effect on linearity [17] and an elegant solution to this problem is still missing.

7.2.3 *Distortion Cancellation Techniques*

Now that the major causes for AM-PM distortion have been highlighted, it is possible to devise circuit techniques to alleviate them, at least to some extend. These techniques aim to (1) linearize the devices, by adding components with a non-linear behavior complementary to that of the active devices used in the PA. (2) De-Qing the input RC product by lowering R_{IN}. (3) Adding harmonic traps, so that the output of a non-linear PA can still resemble the one of a perfectly linear one.

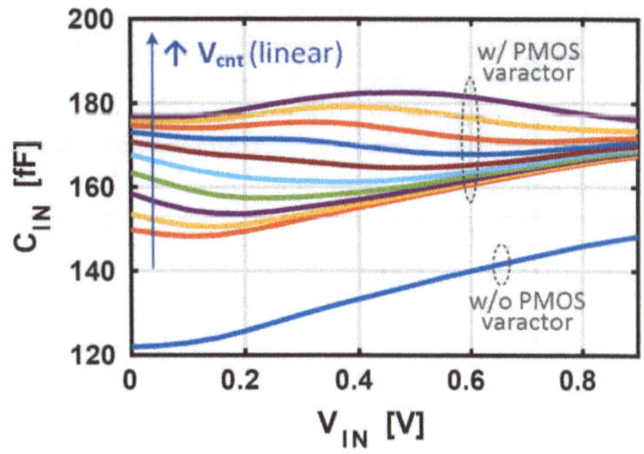

Fig. 7.12 Effect of input PMOS varactors on the input capacitance under large signal. ©2018 IEEE. Reprinted, with permission, from [19]

7.2.3.1 Input PMOS Varactors

The non-linear variation of the input capacitance of a NMOS transistor can be compensated simply by adding a PMOS transistor or (even better) a PMOS varactor at the input (see Fig. 7.7c), as first proposed in [21] and [22] respectively. Figure 7.12 shows C_{IN} versus V_{IN} with and without input varactor. As evident, a PMOS varactor can effectively compensate the non-linear input capacitance of a NMOS CS amplifier. Moreover, the control voltage can be used to compensate for PVT variations or model inaccuracy during the design phase. The total input capacitance is higher, but the equivalent input resistance is much lower due to the limited Q-factor of varactor at mm-Wave. Therefore, the amplifier is also less sensitive to AM-PM conversion coming from an high-Q inter-stage matching network. As it will be shown shortly, this results in a penalty in terms of gain.

The effect on gain and AM-PM when this distortion cancellation technique is applied to the amplifier in Fig. 7.2a is reported in Fig. 7.13. The AM-PM can be effectively compensated when the PA is biased with $I_D/W = 163.4\,\mu A/\mu m$. The 2.2° AM-PM distortion in the case of the classical CS amplifier can be made lower than 1° (see Fig. 7.13e). When the PA is biased deeper in class-AB the befit of this technique is limited. The reduction in term of gain is substantial as shown in Fig. 7.13c, and perhaps not acceptable when the losses of the matching networks at 30 GHz are added. PAE_{1dB} gets closer to PAE_{MAX}, and the benefit in AM-PM is considerable when compared to the 7° of the same amplifier without varactors. However, it is not possible in this case to compensate completely the AM-PM distortion, as clear from Fig. 7.13f. This means that there are other effects that dominates the AM-PM distortion of the PA when biased in deep class-AB. Probably, this has to do with the strong 2nd harmonic component of the output voltage shown in Fig. 7.8f. If this is the case, a 2nd harmonic trap would be more effective.

7.2.3.2 Complementary N-PMOS Amplifier

Another distortion cancellation technique suitable for mm-Wave power amplifiers implemented in deep-scaled CMOS was demonstrated in [20] and its schematic is shown in Fig. 7.14. The combination of NMOS and PMOS common source ampli-

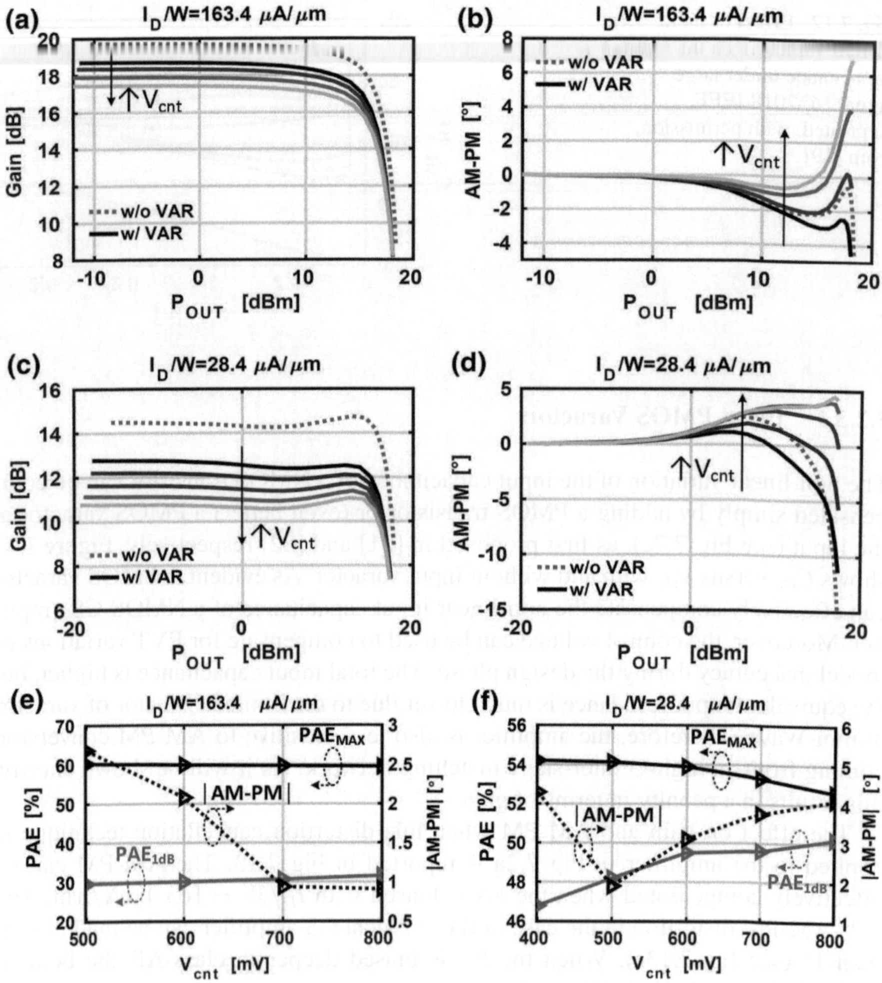

Fig. 7.13 Comparison between different class-AB bias points when input PMOS varactors are added. $I_D/W = 163.4\,\mu\text{A}/\mu\text{m}$ (**a**), (**b**), (**e**) and $I_D/W = 28.4\mu\text{A}/\mu\text{m}$ (**c**), (**d**), (**f**)

fiers, when properly sized and biased, results in C_{GS} distortion cancellation [20] but also in G_m distortion cancellation as LNA designers have learned well in the low-GHz range [23, 24].

This technique is particularly powerful in deep-scaled CMOS since the PMOS devices have benefited more from scaling than NMOS, and today PMOS and NMOS are quite close in terms of g_m and f_t [20, 25]. However, it also have some shortcomings. (1) This technique is not favorable to supply scaling. (2) The PMOS transistor are here used as amplifiers, not just as non-linear passive components for capacitive compensation. Therefore, the modeling at mm-Wave needs to be accurate to correctly design W_P/W_N. (3) The voltage at node V_x is poorly defined. To achieve

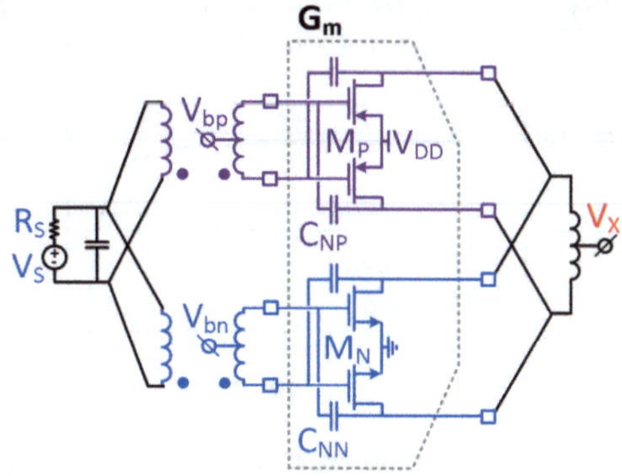

Fig. 7.14 Complementary N-PMOS power amplifier proposed in [20]

optimal operation, the condition $V_x \approx V_{dd}/2$ is need (see Fig. 7.14). This can be achieved under different combinations of V_{bp} and V_{bn}. For given V_{bn}, V_x is quite sensitive to V_{bp}, therefore a feedback loop should be added to this circuit to guarantee proper operation under PVTs, similarly to what proposed in [26] for complementary N-PMOS class-C oscillators.

7.2.3.3 Degeneration Inductance

A degeneration inductance can be added to the CS amplifier for linearization, the resulting circuit is shown in Fig. 7.7d. This technique was first proposed to improve the soft compression behavior of mm-Wave PAs in [27] and then to de-Q the inter-stage matching network and improve AM-PM distortion in [16].

The effect of L_{DEG} is that the 1 dB compression point gets closer to P_{SAT}, therefore $PAE_{1\,\mathrm{dB}}$ gets closer to PAE_{SAT}. The price to pay is a reduced gain [17, 28] (i.e. PAE_{SAT} may be considerably reduced). This could still be acceptable, since in a real application the PA will work far from P_{SAT} to achieve the required specifications in terms of EVM and ACPR.

Another effect of L_{DEG} is that the quality factor of the input impedance of the amplifier significantly reduces [28]. Therefore, the PA is less sensitive to AM-PM distortion coming from the high-Q inter-stage matching [16].

We therefore expect much better AM-AM and AM-PM distortion, improved $PAE_{1\,\mathrm{dB}}$ and excellent performance under modulated signal. At least as long as the transistors are able to provide sufficient gain at the frequency of operation. To verify this point we go back to the design example of Fig. 7.7a. It is worth noting that now both the input matching and the value of C_N need to be adjusted for every L_{DEG} considered. The effect of the degeneration inductance on gain and AM-PM is shown in Fig. 7.15, both when the amplifier is biased with $I_D/W = 163.4\,\mu\mathrm{A}/\mu\mathrm{m}$ and $I_D/W = 28.4\,\mu\mathrm{A}/\mu\mathrm{m}$. In the first case, the added benefit of this linearization technique is not particularly evident. The improvement in terms of $PAE_{1\,\mathrm{dB}}$ and AM-PM distortion is never better than 2.5% and 0.6° respectively (Fig. 7.15e). The advantage

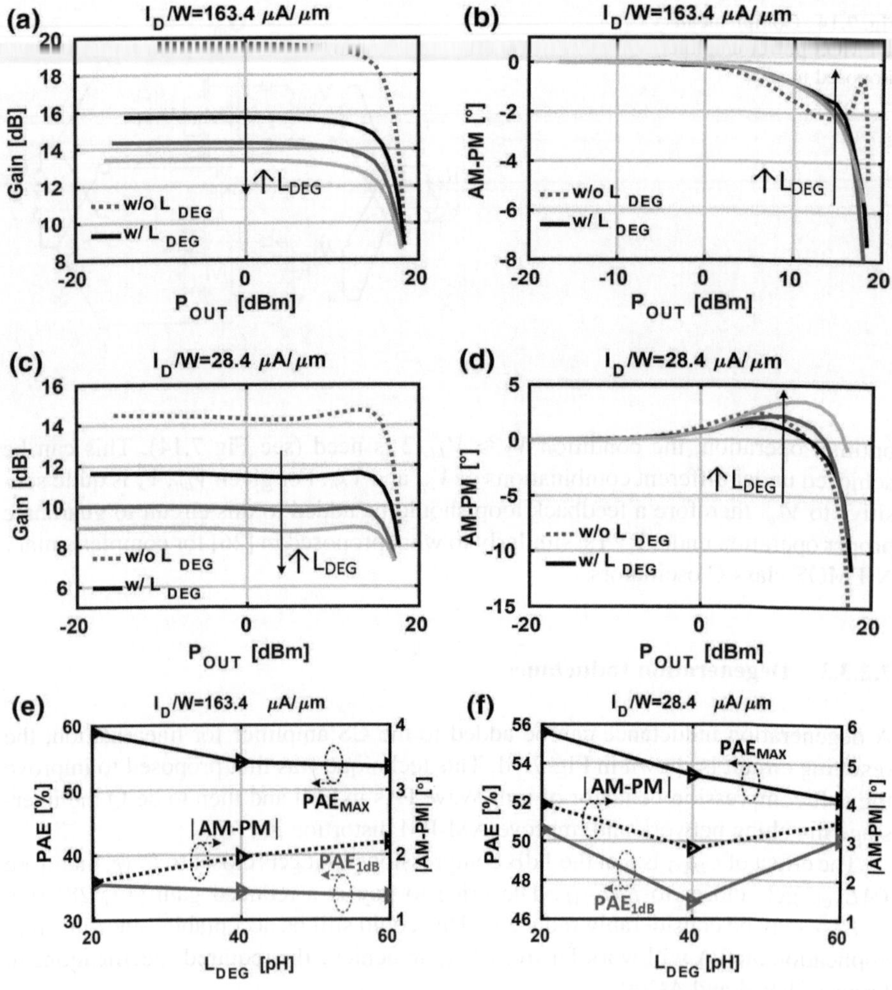

Fig. 7.15 Comparison between different class-AB bias points when a degeneration inductor is added. $I_D/W = 163.4\,\mu A/\mu m$ (**a**), (**b**), (**e**) and $I_D/W = 28.4\,\mu A/\mu m$ (**c**), (**d**), (**f**)

is more evident in deep class-AB, see Fig. 7.15f. Compared to the classical CS amplifier, the AM-PM distortion improves up to ≈4.2°, while still achieving a remarkable $PAE_{1\,dB}$. However, the penalty in gain is also remarkable, and the AM-PM distortion is still not negligible even prior $P_{1\,dB}$. As it will be shown later, PAs that leverage this linearization technique only will still need substantial power backoff to meet the requirements on linearity.

7.2.3.4 Harmonic Traps

As clear from the ongoing discussion, under large-signal conditions a power amplifier biased in deep class-AB shows substantial AM-PM distortion, even when linearization techniques are applied. Moreover, it is well known that the 2nd harmonic component of the output voltage is a key cause of AM-PM distortion [3, 18, 29, 30] and is definitively present at the PA output, see Fig. 7.8f.

The only way to further linearize the PA is therefore introducing harmonic traps, i.e. resonant circuits that shows a low impedance path to ground for the 2nd harmonic current. Three circuit solutions that realize this condition are shown in Fig. 7.16. Before entering the details of each realization, it is worth noting that a harmonic trap will be always effective, no matter what kind of impedance is present at the PA output. This is particularly important in this case, since the amplifier is not well input-out isolated in common mode due to the neutralization capacitor C_N. Moreover, as long as the absolute impedance at the second harmonic is low, we are not really interested in its phase. Therefore, this technique is particularly robust. This is not the case when

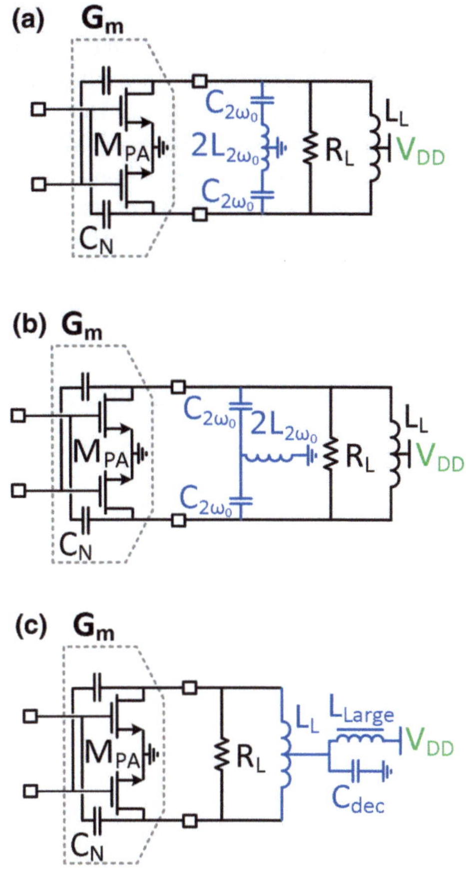

Fig. 7.16 Different possible implementation of 2nd harmonics traps. Classical (**a**), variation proposed in [3] (**b**), solution that takes advantage of the differential nature of the circuit (**c**)

an high impedance termination is needed (as it is the case for class-J or class-F^{-1} operation). In the latter scenario, any low impedance in parallel with the load will impair the effectiveness of the matching network. Moreover, when the absolute value of $|Z_L|$ is large, its phase is of key importance for both PAE and AM-PM distortion [1, 11, 18].

The solution in Fig. 7.16a introduces single-ended series LC filters to realize a short at the 2nd harmonic. The lower the value of the series capacitance, the higher the Q of the filter. Clearly, the quality factor of the filter impose a trade-off between the maximum achievable 2nd harmonic suppression and the bandwidth of the filter. Moreover, the harmonic traps in Fig. 7.16a are present both in common mode and in differential mode. Due to the limited quality factor of a practical on-chip implementation, the effect on the load impedance at the fundamental frequency will not be negligible, further complicating the design. A slightly better solution has been proposed in [3] and is shown in Fig. 7.16b. In this case only the capacitors appear in DM, while the losses of the inductor do not effect, at least in theory, the efficiency at the first harmonic. However, (1) capacitors are particularly lossy at mm-Wave. And (2) the output capacitance of a power amplifier is normally already large. Such a large $C_o + C_{2\omega_o}$ will necessitate an even lower inductance value to resonate at the fundamental frequency. This may result in further degradation of the insertion loss of the filter, since the Q-factor of inductors is not constant with L and drops when an excessively low inductance value is required [31]. A more elegant solution is shown in Fig. 7.16c. Here the harmonic trap affects the CM only, not degrading the impedance seen in DM and therefore efficiency. Further, we note that the inductance provided by a bond wire $\approx 1nH$ is already sufficient to mimic an RF choke at 2 × 30 GHz. By properly design $C_{dec} = 1/((2\pi 2f_o)^2 L_L)$ an effective harmonic trap can be designed. This simple example shows the importance of the proper design of the bias network for a power amplifier. A state-of-the-art design that leverage to some extend this technique and shows its limitation when a broadband PA is designed is reported in [22].

The harmonic trap in Fig. 7.16c has been embedded in the previously designed PA biased in deep class-AB. Figure 7.17 shows the resulting effect on gain and

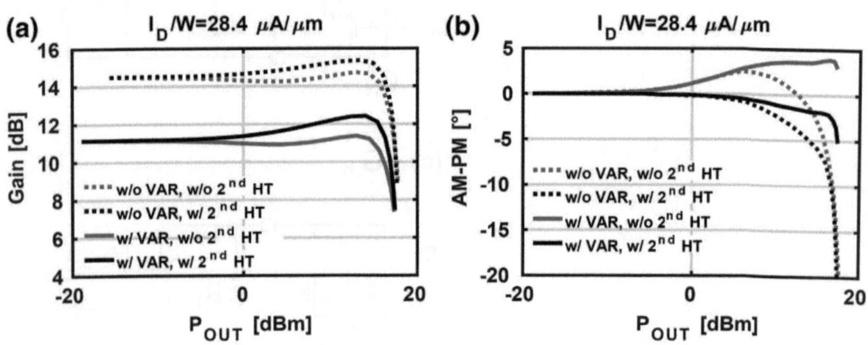

Fig. 7.17 Effect of 2nd harmonic short with (continuous line) and without (dotted line) input varactors on gain (**a**) and AM-PM distortion (**b**)

AM-PM distortion both in the classical case and when PMOS varactors are added. The combination of the 2nd harmonic trap and input varactors results in less than 2.7° AM-PM distortion, scc Fig. 7.17b. It is worth noting that the gain expansion is now enhanced, therefore the PA could be designed at higher quiescent current. This would result in overall higher small signal gain and lower AM-PM distortion. This simple design example shows the potential of these techniques. However, the inherent trade-off between 2nd harmonic suppression and achievable bandwidth still poses severe doubts on the effectiveness of such techniques when a modulated signal with large RF bandwidth is to be transmitted.

7.3 Design Example: A Highly Linear Wideband PA in 28nm CMOS

Figure 7.18 shows the mm-Wave frequency bands allocated in different part of the world for future 5G wireless communication systems. In this scenario, a mm-Wave massive MIMO/phased array system would greatly benefit from a single high-performance wideband PA. Moreover, as clear from the discussion above, techniques to effectively compensate AM-PM distortion over the whole bandwidth of operation are needed as well, to improve both in-band and out-of-band linearity (i.e. EVM and ACPR).

Recently, the work in [16] has shown the feasibility of 28 GHz CMOS PAs with outstanding efficiency. However, the BW, average P_{OUT} and ACPR under modulated signal considerably limit the achievable link distance and coexistence with adjacent

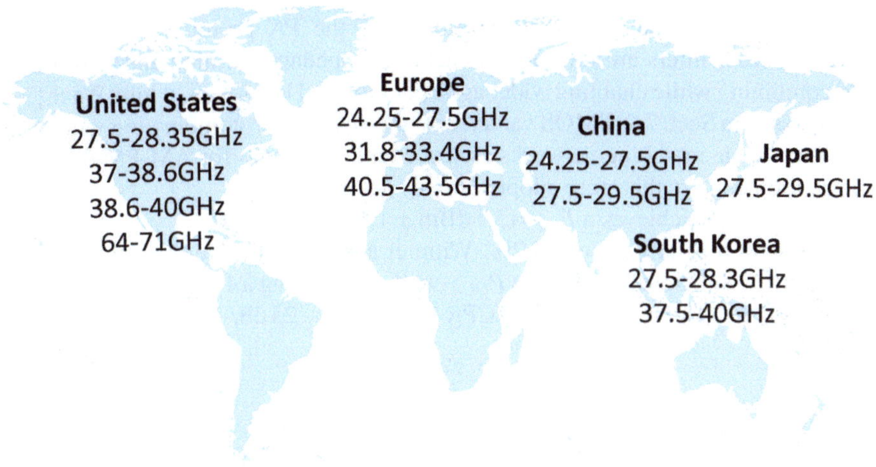

Fig. 7.18 Frequency bands allocated in different part of the world for future 5G wireless communication systems [32]

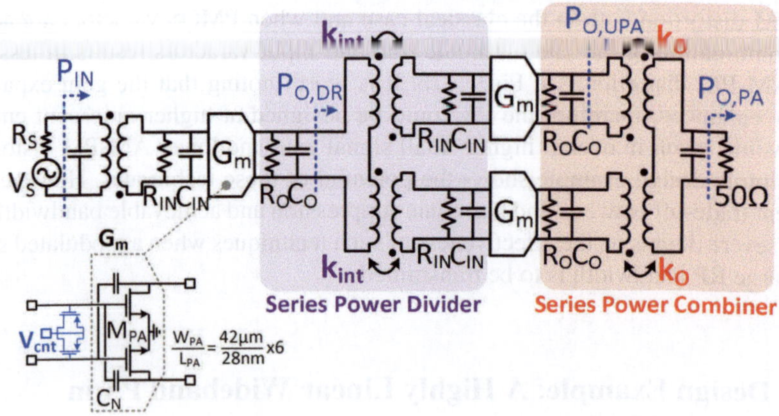

Fig. 7.19 Simplified schematic of the proposed power amplifier with wideband AM-PM linearization. ©2018 IEEE. Reprinted, with permission, from [19]

channels. 2nd harmonic shorts are introduced in [3] to enhance the PA linearity, achieving excellent PAE under modulated signal but with very limited bandwidth. Clearly, the effectiveness of harmonics traps increase with their quality factor, directly trading with bandwidth. In [20] a complementary N-PMOS PA to cancel the AM-PM distortion due to the efficient class-AB operation is reported, but this approach is not favorable to supply scaling.

This section presents a 29–57 GHz (65% BW) class-AB PA tailored for 5G phased arrays, designed in 0.9 V 28 nm bulk CMOS, without RF thick top metal. The trade-offs between bandwidth and power added efficiency are discussed in great details. Second order effects due to physical layout implementation are also addressed. Figure 7.19 shows the simplified schematic of the PA prototype. Transformer-based 4th order filters are leveraged to achieve impedance scaling and power division/combining while enabling wideband operation and low losses at mm-Wave [33]. As discussed in Sect. 7.2, PMOS varactors at the G_m input in combination with wide-band inter-stage matching networks are excellent candidates for AM-PM distortion cancellation and are therefore adopted in this design.

The prototype achieves a P_{sat}=15.1 dBm ± 1.6 dB and |AMPM| < 1° from 29 to 57 GHz, with a peak PAE of 24.2%. Without applying any pre-distortion, the PA delivers 10.1, 8.9, 5.9 dBm average P_{OUT} while amplifying a 1.5, 3, 6 Gb/s 64-QAM respectively at 34 GHz with EVM/ACPR better than −25 dB/−30 dBc.

7.3.1 Broadband Impedance Transformation

To deliver the required linear output power while operating at 0.9 V nominal supply the 50 Ω load impedance needs to be scaled down. Figure 7.20a shows the proposed

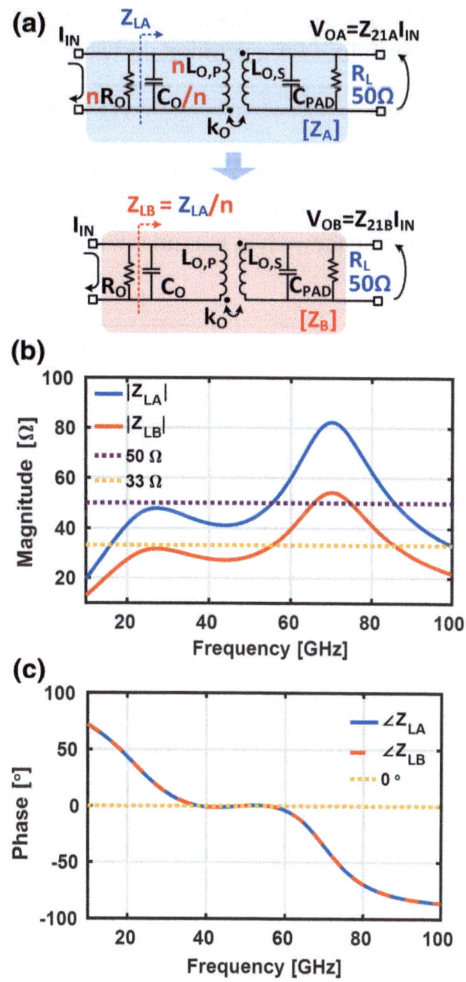

Fig. 7.20 **a** Broadband transformer-based 4th order filter and circuit transformation to realize $1/n$ impedance scaling without impairing the frequency response. Magnitude (**b**) and phase (**c**) of the load impedance presented at the PA output versus frequency with and without impedance scaling. ©2018 IEEE. Reprinted, with permission, from [19]

circuit transformation adopted to absorb the layout parasitics of the transconductors and the output pads while achieving the required $1/n$ impedance scaling at the cost of a $1/\sqrt{n}$ reduction of the transimpedance gain Z_{21}. By taking advantage of the properties of the transformer, this 4th order filter can realize broadband impedance transformation without adding any lossy component.

The center frequency of the filter in Fig. 7.20a can be written as

$$f_o = \frac{1}{2\pi\sqrt{L_{O,P}(1-k_O^2)C_O}} = \frac{1}{2\pi\sqrt{L_{O,S}(1-k_O^2)C_{PAD}}}. \tag{7.5}$$

The kQ product of this filter at f_o can be defined as (the interested reader is referred to Appendix I for a detailed derivation)

$$kQ = |k_O|\sqrt{Q_O\,Q_{load}}, \tag{7.6}$$

where $Q_O=2\pi f_o R_O C_O$, $Q_{load} = 2\pi f_o R_L C_{PAD}$ and $R_L = 50\,\Omega$ in this example.

When the kQ product of the filter is reasonably high, the load impedance at the PA output shows two maxima in the proximity of the two resonant frequencies [33]

$$f_L = \frac{1}{2\pi\sqrt{L_{O,P}(1+|k_O|)C_O}} = \frac{1}{2\pi\sqrt{L_{O,S}(1+|k_O|)C_{PAD}}}, \tag{7.7}$$

$$f_H = \frac{1}{2\pi\sqrt{L_{O,P}(1-|k_O|)C_O}} = \frac{1}{2\pi\sqrt{L_{O,S}(1-|k_O|)C_{PAD}}}. \tag{7.8}$$

Most importantly, as calculated in Appendix II, the real part of the load impedance presented at the PA output at the resonant frequencies f_L and f_H can be written as

$$Re\{Z_{LA}(f_{L,H})\} = \frac{n\,L_{O,P}}{L_{O,S}}R_L = \frac{n\,C_{PAD}}{C_O}R_L. \tag{7.9}$$

From Eq. 7.9, when $C_O/n = C_{PAD}$, $Re\{Z_{LA}(f_{L,H})\} = R_L$. Figure 7.20b, c show the simulated magnitude and phase of the impedance at the PA output when $C_O = 116$fF, $C_{PAD} = 65$fF, $k_O = 0.8$ and $n = 50/33\,\Omega$. The results presented here are equivalent to the ones presented in [34]. However, the circuit techniques proposed in this work are remarkably simpler and do not require a Norton transformation.[1] As a result, the effect of the transformer design parameters on the frequency response of the filter are immediately evident.

It is worth mentioning that in the design example considered so far the losses of the transformer have been neglected, with the aim of providing intuition on the operation of the circuit. As it will be shown in the following, the insertion loss of the filter does impact the frequency response, significantly attenuating the second resonant peak shown in Fig. 7.20b.

7.3.2 Transformer-Based Output Combiner and Inter-stage Power Divider

To achieve the required high power levels while operating at a nominal power supply below 1V, CMOS power amplifiers largely relies on power combining techniques

[1] The work presented in [34] proposes a four-step procedure that relies on four design parameters (i.e., d, m, l, and n) and Norton transformation to derive the design parameters of the final transformer-based 4th order filter from inductively coupled resonators.

[13, 15, 35–37]. The 2-port magnetically coupled resonator shown in Fig. 7.20a can be extended to realize 3-port power combiners by applying the circuit transformation in Fig. 7.21a. To compensate for the parasitic inductance of the interconnections to the RF pads (L_{PAD}), the magnetic coupling coefficient needs to increase and the self-inductance of the secondary winding needs to be sized down (see Fig. 7.21a). This clearly shows the importance of reducing L_{PAD} by proper layout. Similarly, a 3-port power divider can be derived by simply inverting the input and output ports [33]. However, in this case no impedance scaling is needed and the higher Q-factor of the load ($R_{IN}C_{IN} > R_O C_O$) imposes $k_{int} < k_O$, limiting the BW of the PA for a given in-band ripple [33]. Figure 7.21) shows the post-layout simulations of the gain from the driver input to the input of the unit PA and to the load. It is remarkable that the frequency response of the interstage power divider shows clearly the two maxima expected from theory, even if in any practical realization the transformer properties (i.e. magnetic coupling coefficient, self inductances and quality factors) vary with frequency, to some extent. This one again proves the strength of the theoretical analysis provided in this work, as well as in prior art [32–34, 38, 39].

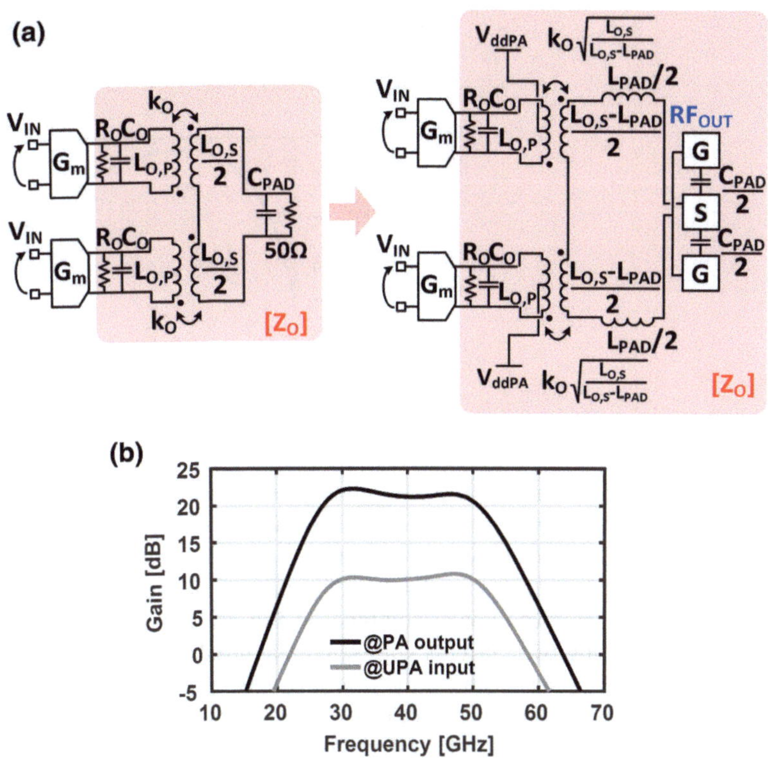

Fig. 7.21 **a** Circuit transformation to realized a 3-port broadband series power combiner and include the layout parasitics of the interconnections to the GSG probe-pads (L_{PAD}, C_{PAD}). **b** Simulated gain from the input of the driver to the input of the unit PA (UPA) stage (gray line) and simulated gain of the full PA (black line). ©2018 IEEE. Reprinted, with permission, from [19]

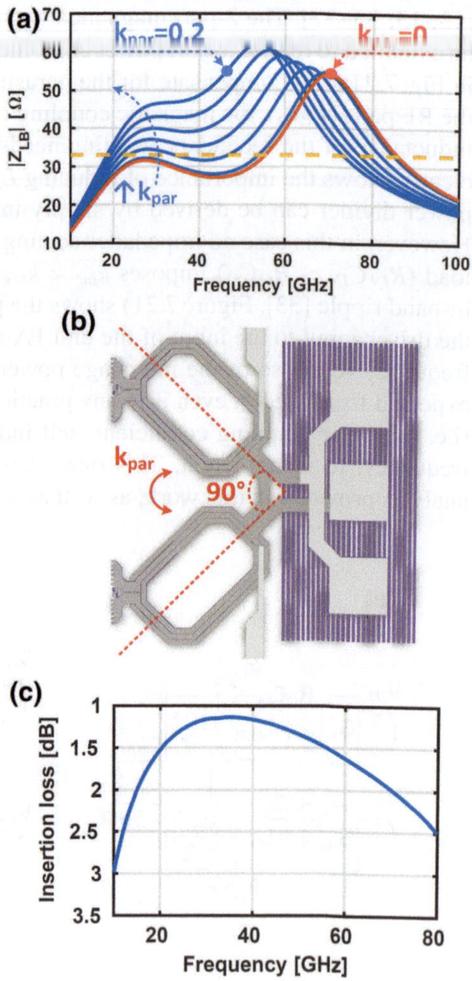

Fig. 7.22 **a** Detrimental effect of the parasitic magnetic coupling between the two transformers of the series power combiner (k_{par}) on the load impedance and **b** adopted layout to minimize it. k_{par} is swept linearly from 0.01 to 0.2 in 6 steps (blue line). $k_{par} = 0$ is also added in red line for comparison. **c** Simulated combiner insertion loss versus frequency. ©2018 IEEE. Reprinted, with permission, from [19]

Although power combiners and power dividers seem to be linked by the same theory except for a simple inversion of the input/output ports, there is a key difference that needs to be clarified. In a power divider there is only one port delivering power while the other two output ports are loaded by a passive network. This is not the case in power combiners. The power combiner in Fig. 7.21a assumes that the two input ports are driven by two currents with same magnitude and phase. When this assumption is not valid, the analysis of the resulting network is much more involved [14, 37, 40–44] and beyond the scope of this work.

Figure 7.22a shows a second order effect often neglected in a power combiner, the parasitic magnetic coupling between the two series transformers (k_{par}). When k_{par} increases from the ideal 0–0.2, the synthesized impedance rises significantly over the whole band. To minimize this detrimental effect, the layout in Fig. 7.22b is adopted in this design. Compared to the distributed active transformer (DAT) topology adopted in [15, 20, 35, 38], this solution allows more freedom in the layout while significantly reducing k_{par} at the expenses of higher losses (from post layout EM simulations ≈0.14 dB higher insertion loss in this design). Relative to a layout implementation where the two series transformers are laid out parallel to each other with an edge-to-edge spacing of 10 μm, the adopted solution allows to reduce L_{PAD}, resulting in ≈0.26 dB lower insertion loss, while achieving lower k_{par}. The simulated k_{par} of the combiner in Fig. 7.22b is less than 0.01 at 40 GHz. The minimum insertion loss is −1.14 dB at 35 GHz with a $BW_{-1\,\mathrm{dB}}$ from 14 to 73 GHz as shown in Fig. 7.22c.

It is worth noting that as the minimum feature size scales, every technology node imposes increasingly strict design rules. Higher metal densities are to be fulfilled in smaller area windows. As a result, mm-Wave on-chip transformers suffers from higher insertion loss. Whereas the effect of metal dummies on the magnetic coupling coefficient, primary and secondary inductance is negligible [13, 45]. Moreover, to comply to the design rule check (DRC) the device gates need to be laid out with the same orientation across the whole wafer. From the point of view of the designer, this means that the layout of the power transistor optimized for mm-Wave operation (e.g. the one proposed in [13]) can not be rotated 90° as it used to be the case in less advanced technology nodes. The transformer layout adopted in this work for the power combiner allows to keep the same orientation for both the power amplifier and the driver stage, fulfilling this important task. This is not straightforward if the power combiner is implemented with a DAT [15, 20, 38].

Noteworthy, transmission line based matching networks are also suitable for broadband PA design [14, 44, 46]. These networks typically show lower losses when compared to lumped element component implementations. However, transformer based filters enable lower silicon area consumption while providing galvanic isolation, protection against ESD events, balance to unbalance conversion and ease the DC bias feed to the active circuitry. Therefore, in this work lumped element components were preferred.

7.3.3 More on the kQ Product

In the following we will investigate the effect of the kQ product on (1) the frequency response of the filter, (2) on the transformer insertion loss and (3) the trade-off between $BW_{-3\,\mathrm{dB}}$ and power added efficiency in mm-Wave power amplifiers that leverage transformer-based 4th order matching networks.

So far we have assumed that the kQ product of the filter terminations shown in Fig. 7.20a is high enough. However, the 50 Ω load at the PA output and the limited practical values of C_{PAD}, C_O significantly challenge the aforementioned assumption, as clear from Eq. 7.6. Although the circuit transformation in Fig. 7.20a guarantees $Z_{LB} = Z_{LA}/n$ without impairing the frequency response of the filter, the magnetic coupling coefficient of the output transformer (k_O) needs to be designed as large as possible to synthesize the required impedance over the $BW_{-3\,\mathrm{dB}}$ of the PA. This effect

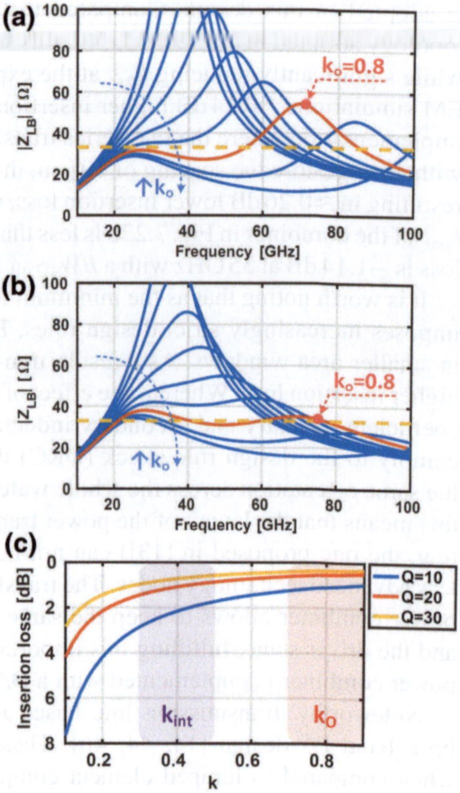

Fig. 7.23 Simulated effect of a limited magnetic coupling k_O on the load impedance when a lossless (**a**) and lossy (**b**) transformer model is used. k_O is swept from 0.5 to 0.9 in steps of 0.05. **c** Simulated insertion loss of a transformer against magnetic coupling coefficient k for different Q-factors. ©2018 IEEE. Reprinted, with permission, from [19]

is clearly shown in Fig. 7.23a. In this design example, f_L in Eq. 7.7 is kept constant to 25 GHz, and k_o is swept, changing the position of the second resonant frequency of the filter. By rearranging Eqs. 7.7 and 7.8, f_H can be expressed as

$$f_H = f_L \sqrt{\frac{1 + |k_O|}{1 - |k_O|}}. \tag{7.10}$$

Let us now consider a limited quality factor for the transformer primary and secondary inductors ($L_{O,P}$ and $L_{O,S}$ in Fig. 7.20a) and model the losses with series resistors, $R_{L_{O,P}} = \omega L_{O,P}/Q_P$ and $R_{L_{O,S}} = \omega L_{O,S}/Q_S$ respectively [47]. Intuitively we expect that since the quality factor of the filter is impaired by the transformer losses, to keep the kQ product high enough k_O needs to increase accordingly. Figure 7.23b shows the effect on the load impedance when the transformer previously designed exhibits a $Q_P = Q_S = 10$ at 30 GHz. The magnitude of the load impedance at the first resonant frequency f_L is slightly increased, while a the second resonant peak f_H it is highly attenuated. Indeed if the kQ product is not high enough it is not possible to realize the required impedance transformation.

Another key aspect for a matching network at the PA output is its insertion loss. When a lossy transformer is used in the filter, we expect that a higher magnetic coupling factor k results in larger induced current at the secondary coil, leading to lower insertion loss. This is true as long as the quality factor of the network does not change. The insertion loss of a lossy transformer can be expressed as [48, 49]

$$PL_{trasf} = 20log_{10}\left[\frac{kQ}{-1+\sqrt{1+(kQ)^2}}\right], \tag{7.11}$$

where in this case $Q = \sqrt{Q_P Q_S}$, and Q_P, Q_S are the quality factors of the transformer primary and secondary windings. PL_{trasf} against the magnetic coupling coefficient for different Q is reported in Fig. 7.23c. Indeed, the lower k the higher the insertion loss. However, if the quality factor is high enough, the degradation of the transformer performance is limited.

To put the foregoing discussion in perspective and get a deeper understanding on the trade-off between PAE and bandwidth for a mm-Wave power amplifier, let us consider the following design example. The two-stage PA in Fig. 7.19 is designed to deliver a linear output power $P_{O,PA}$ equal to 14 dBm to the 50 Ω load. A power consumption of $P_{dc,PA} = 120\,\text{mW}$, $P_{dc,DR} = 50\,\text{mW}$ and a gain of $G_{U,PA} = 10\,\text{dB}$, $G_{DR} = 17\,\text{dB}$ for the PA and driver stage respectively are considered. Further, 5 dB insertion loss is accounted for the input matching network (PL_{in}), due to poor matching to 50 Ω or to explicit resistors added to reduce the RC product at the input of the G_m stage (e.g. as in [9, 50]). The output combiner is designed with high k_O to ensure low losses. In the following we will assume $PL_{comb} = 1\,\text{dB}$, a reasonable value for a 28 nm CMOS process without ultra-thick top metal, see Fig. 7.22c. Due to the low Q of the filter termination, a very broadband impedance transformation can be achieved, as shown in Fig. 7.23a. Since the input impedance of the G_m stage shows a much higher RC product, the bandwidth of the PA will be limited by the frequency response of the inter-stage power divider. A higher k_{int} allows a larger bandwidth at the expenses of larger in-band ripple. In the limit case of $k_{int} = k_O = 0.8$, f_H in Eq. 7.10 is pushed to $3f_L$ and the in-band ripple is so large that the resulting $BW_{-3\,\text{dB}}$ of the full PA is narrow (typically <15% [16]). For these reasons broadband amplifiers often adopt a moderate $k_{int} \approx 0.4$ [33, 38, 50]. From Eq. 7.11, the IL_{trasf} when Q is equal to 30, 20, 10 and $k_{int} = 0.4$ is respectively 0.4, 0.5, and 1.1 dB larger than the insertion loss in the case of $k_O = 0.8$. In the worst case scenario of Q = 10 and referring to Fig. 7.19, the PAE for the complete PA can be calculated as

$$PAE = \frac{P_{O,PA,lin} - P_{IN,lin}}{P_{dc,DR} + P_{dc,PA}}, \tag{7.12}$$

where $P_{O,PA,lin}$ and $P_{IN,lin}$ are expressed in watt and $P_{IN} = P_{O,PA} + PL_{comb} - G_{U,PA} + PL_{div} - G_{DR} + PL_{in}$ (see Fig. 7.19). The power added efficiency in Eq. 7.12 is 14.6% in the narrowband case ($k_{int} = k_O = 0.8$, $PL_{div} = 1.8\text{dB}$), and degrades only 0.04% when a broadband inter-stage power divider is used ($k_{int} = 0.4$, $PL_{div} = 2.15\,\text{dB}$).

Clearly, benefited by the high g_m and f_t available in 28 nm CMOS, the design trade-off between bandwidth and linearity for a mm-Wave PA is particularly relaxed. We further note that out of band emissions at the 2nd harmonic are heavily suppressed by the differential operation of the PA, while out of band emissions at the 3rd harmonic fall at 3× higher frequencies and will experience substantial filtering after the PA by the antenna. Considering the improvements in terms of AM-PM and memory effects compared to a narrowband counterpart discussed, a wideband design is preferable.

7.3.4 Measurement Results

Figure 7.24 shows the die picture of the PA prototype realized in 28 nm bulk CMOS without ultra-thick top metal [51]. The core silicon area is 0.160 mm^2, including the input/output RF pads. Measurements are performed on a high frequency probe station. The DC pads are wire-bonded to a PCB while the RF input and output pads are accessed by GSG probes. In this design the 0.9 V nominal supply voltage of this technology is used. The PA and the driver are biased in class-AB with bias current densities of 135 and 90 μA/μm respectively. The control voltage of the varactors is set to $V_{cnt} = 0.7$V (unless otherwise stated).

7.3.4.1 CW Measurements

Figure 7.25a shows the measured S-parameters. The PA achieves 20.8 dB gain over a 29–57 GHz (65%) $BW_{-3\,\mathrm{dB}}$. The input is not matched to 50 Ω, resulting in a lower power delivered to the PA. However, the high gain (≈20 dB) limits the impact on

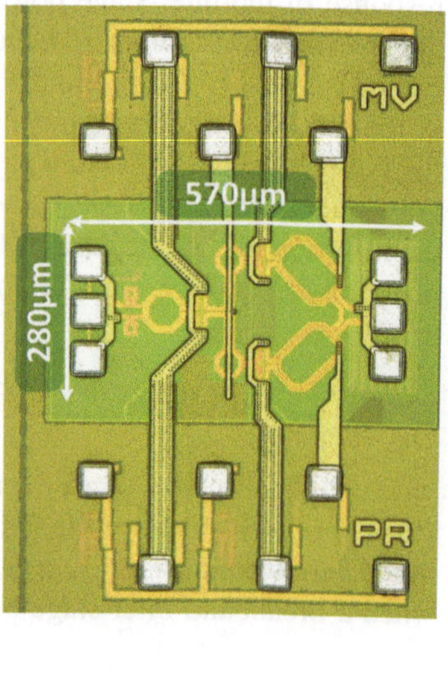

Fig. 7.24 Die picture of the realized 28 nm CMOS prototype. ©2018 IEEE. Reprinted, with permission, from [19]

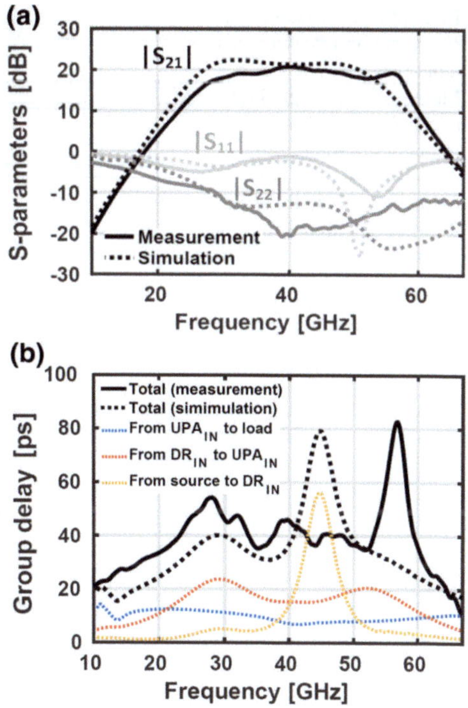

Fig. 7.25 **a** Measured (continuous line) and simulated (dotted line) S-parameters versus frequency. **b** Measured (continuous line) and simulated (dotted line) group delay versus frequency. ©2018 IEEE. Reprinted, with permission, from [19]

PAE. The PA is unconditionally stable from DC to 67GHz. The measured group delay shows less than ±10.3ps variation from 20 to 54 GHz, see Fig. 7.25b. This constant group delay together with the flat in-band gain response is key to amplify broadband modulated signals without degrading EVM [15, 39]. Figure 7.25b shows also the simulated group delay stage by stage. The group delay of the interstage power divider dominates the frequency response at the first peak, while the group delay of the input matching network dominates in the higher part of the spectrum (see Fig. 7.25b).

Figure 7.26a shows the measured large-signal continuous-wave (CW) performance against P_{IN} at 30G Hz. Figure 7.26b reports the measured output power and PAE performance against frequency. The measured peak P_{sat} is 16.6 dBm, P_{1dB} is 13.4 dBm, PAE_{MAX} is 24.2% and $PAE_{1\,\mathrm{dB}}$ is 12.6%. The DC power consumption of the final stages of the PA and the driver is 137 and 33.1 mW respectively at PAE_{MAX}. The P_{sat} and $P_{1\,\mathrm{dB}}$ $BW_{-1\,\mathrm{dB}}$ are 56.6 and 32.3%. The PAE_{MAX} and $PAE_{1\,\mathrm{dB}}$ $BW_{-1\,\mathrm{dB}}$ are 39.5 and 32.3% respectively. The efficiency of the PA is reduced in the higher part of the spectrum due to the insertion losses of the output combiner (see Figs. 7.22c and 7.26b).

The AM-PM is measured with an Agilent N2447A PNA-X network analyzer as in [20]. The measured |AM-PM| distortion at $P_{1\,\mathrm{dB}}$ is less than $<1.8°$ from 26 to 58 GHz and can be further reduced to $<1°$ by fine tuning the control voltage of the varactors V_{cnt} from 0.5 to 0.9 V as shown in Fig. 7.26c. In the latter case V_{cnt} is tuned

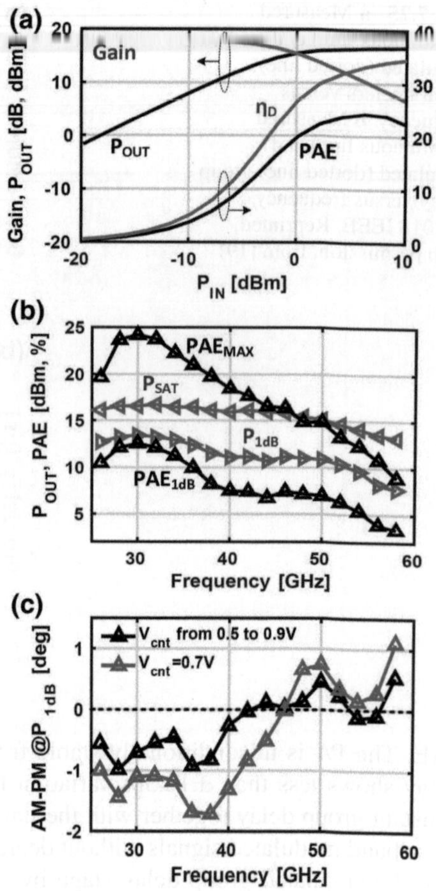

Fig. 7.26 **a** Measured gain, P_{OUT}, η_D and PAE against input power at 30GHz. **b** Measured large-signal CW performance versus frequency. **c** Measured AM-PM at P_{1dB} versus frequency. ©2018 IEEE. Reprinted, with permission, from [19]

at each frequency for minimum AM-PM. The measured AM-PM versus P_{OUT} at 25 and 56 GHz when V_{cnt}=700 mV is shown in Fig. 7.27a. To further verify the effect of the varactor control voltage on the operation of the PA, the measured and AM-PM versus output power at 44 GHz for V_{cnt} that varies from 500 to 900 mV is shown in Fig. 7.27b. The effect of V_{cnt} on AM-AM is limited, while always ensuring excellent AM-PM linearity, proving the robustness of this technique. Further, the frequency response of the amplifier is almost insensitive to V_{cnt} up to the center frequency as shown in Fig. 7.27c. The gain variation is more pronounced at higher frequencies, limiting the practical AM-PM fine compensation through V_{cnt} in this part of the spectrum. This can be intuitively explained by referring to Fig. 7.12b. The variation of the input capacitance of the transconductance stage is very limited over a large range of V_{cnt} and V_{IN}. However, the varactor Q degrades with frequency resulting in higher losses in this part of the spectrum.

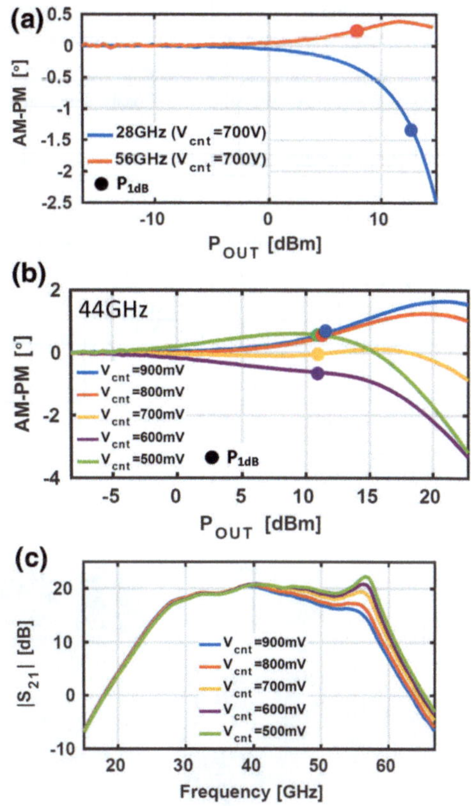

Fig. 7.27 **a** Measured and AM-PM versus output power at 25 GHz and 56GHz (V_{cnt}=700 mV). **b** Measured and AM-PM versus P_{OUT} at 44 GHz when V_{cnt} is varied from 500 to 900 mV. **c** Measured $|S_{21}|$ versus frequency when V_{cnt} is varied from 500 to 900 mV. ©2018 IEEE. Reprinted, with permission, from [19]

7.3.4.2 Modulated Signal Measurements

The PA was tested applying a 64-QAM modulated signal with 0.35 roll-off factor raised-cosine shaped filter and 8.3 dB PAPR. The block diagram of the full measurement setup used for modulated signal measurements is reported in Fig. 7.28. The power at the input and output of the DUT is measured with wideband power meters, care is taken to suppress the LO feed-through and accurately de-embed the losses. The EVM is measured with a high frequency oscilloscope with built-in VSA software. The ACPR is measured with a 43.5 GHz R&S FSW and the DC power consumption is measured with two accurate source meters and a power analyzer.

Figure 7.29a shows the measured constellation and EVM summary at 34 GHz with 1.5 Gb/s data rate and 10.1 dBm average P_{OUT}. In this work, EVM is normalized to the reference RMS power ($\approx$3.7 dB worse than when normalized to the constellation maximum for a 64-QAM) [52]. The ACPR for the same signal is shown in Fig. 7.29b. Figure 7.29c shows the measured P_{OUT} and PAE at 1.5, 3 and 6 Gb/s data rate from 26 to 34 GHz at EVM< -25 dB.

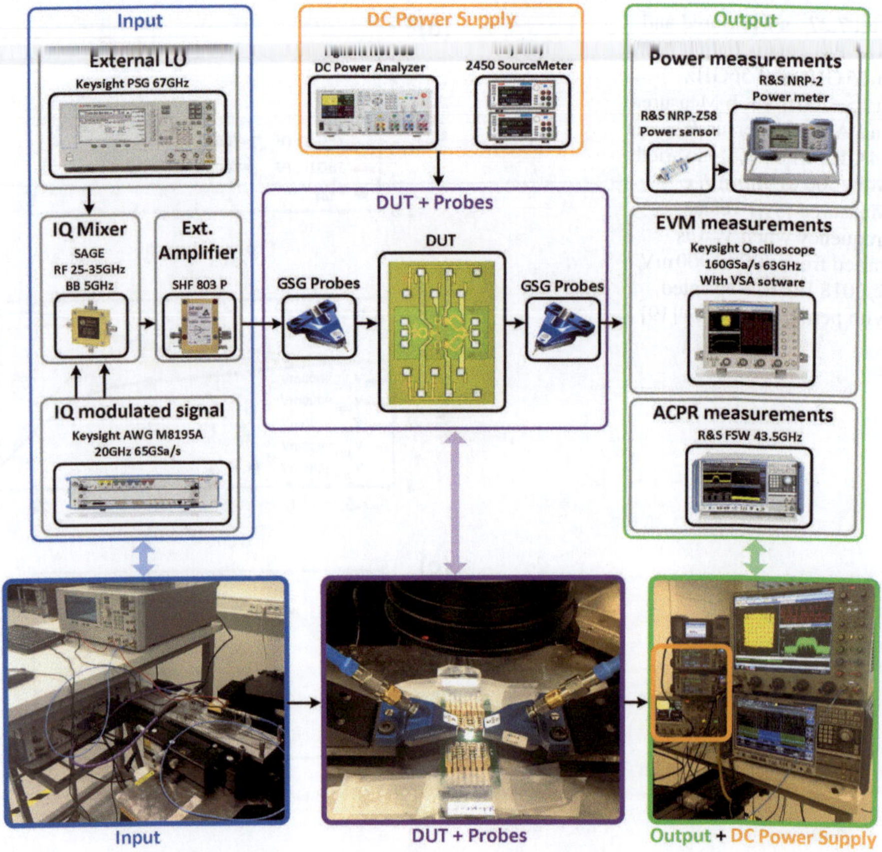

Fig. 7.28 Block diagram of the modulated signal measurement setup

Figure 7.30 shows the measured upper and lower ACPR from 26 to 34 GHz at EVM<-25 dB under 64-QAM signal when the modulation bandwidth (data rate) is increased from 0.337 GHz (1.5 Gb/s) to 0.675 GHz (3 Gb/s) and 1.35 GHz (6 Gb/s). Benefited by the discussed design techniques the prototype shows a measured ACPR always better than −30 dBc while the difference between upper and lower ACPR never exceeds 2.2 dB. This demonstrates the excellent out-of-band linearity and very limited memory effect of the proposed power amplifier.

Although modulated signal measurements beyond 34 GHz were limited by the setup, the continuous wave measurements in Fig. 7.26a, b suggest that good linearity could be achieved also at higher frequencies at the cost of reduced power added efficiency.

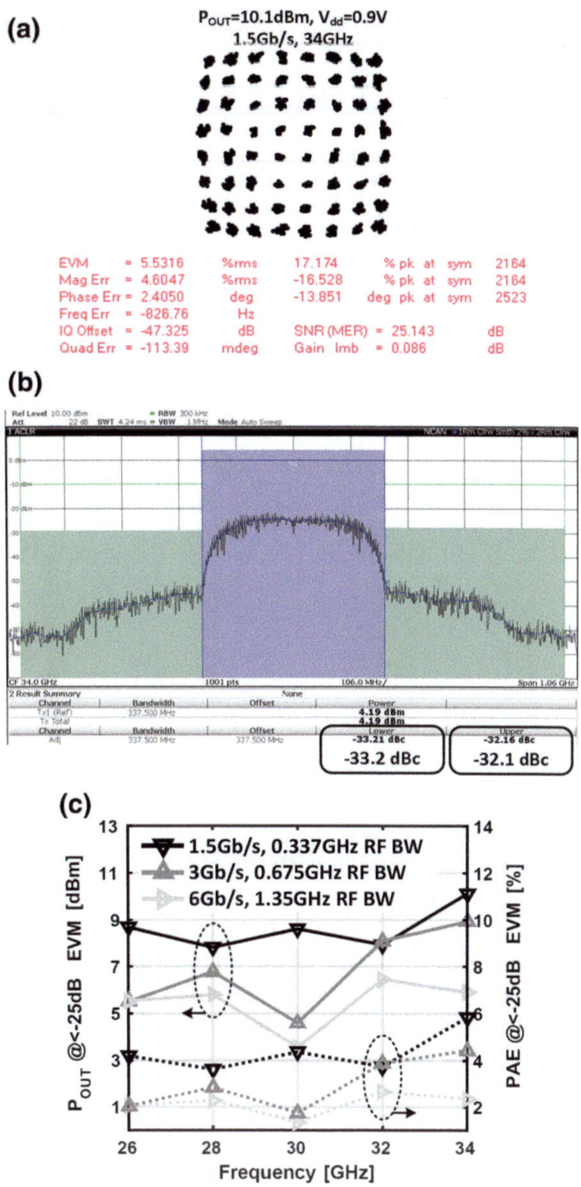

Fig. 7.29 Measured constellation and EVM summary (**a**), and ACPR (**b**) of a 1.5 Gb/s data rate 64-QAM modulated signal at 34 GHz/10.1 dBm f_c/P_{OUT}. **c** Measured P_{OUT} (continuous line) and PAE (dotted line) at EVM<-25 dB for a 64-QAM at 1.5, 3, 6 Gb/s data rate versus frequency. ©2018 IEEE. Reprinted, with permission, from [19]

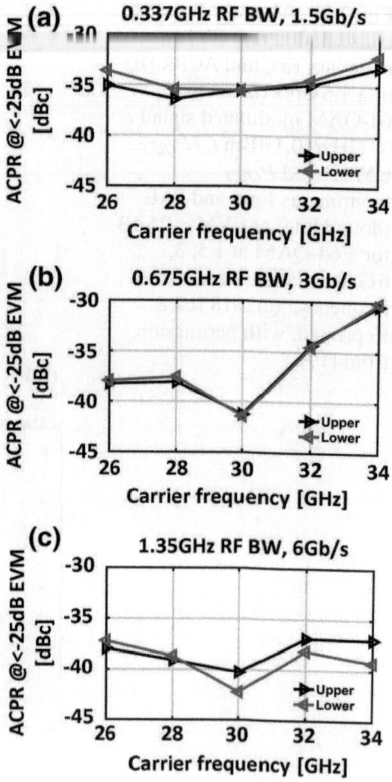

Fig. 7.30 Measured upper (black line) and lower (gray line) ACPR at EVM<-25 dB for **a** 64-QAM at 1.5 Gb/s, **b** 3 Gb/s, **c** 6 Gb/s data rate versus frequency. ©2018 IEEE. Reprinted, with permission, from [19]

7.3.4.3 Comparison with the State-of-the-Art

Tables 7.1 and 7.2 summarize the measured results and provide a comparison with recently published state-of-the-art mm-Wave PAs realized in silicon-based technologies. The presented design shows the widest reported fractional S_{21} $BW_{-3\,\mathrm{dB}}$ and P_{sat} $BW_{-1\,\mathrm{dB}}$ while demonstrating excellent AM-PM linearity over the whole band of operation and still achieving a remarkable 24.2% peak PAE_{MAX} over a 39.5% $BW_{-1\,\mathrm{dB}}$.

The advantages of the presented circuit techniques stand out when a modulated signal with large bandwidth is applied. For the same data rate and EVM specifications, this PA delivers 5.9 and 4.7 dB higher average P_{OUT} than [16] and [20] respectively, despite the lower V_{DD}. The PA in [3] that leverages 2nd harmonic traps achieves higher P_{OUT} under modulated signal but with a much higher (×2.4) supply voltage and much lower modulation bandwidth (less than 32%). The 0.13 μm SiGe PA presented in [32] shows excellent efficiency under modulated signal, however it does not achieve low AM-PM distortion over the whole $BW_{-3\,\mathrm{dB}}$ and the varactor-loaded

Table 7.1 Comparison with state-of-the-art silicon-based PAs, CW performance. ©2018 IEEE. Reprinted, with permission, from [19]

Ref.	**This work** [19]	ISSCC17 [32]	RFIC17 [44]	JSSC17 [14]	ISSCC17 [50]	JSSC16 [16]	TMTT16 [3]	TMTT16 [3]	TMTT16 [20]	CICC15 [38]	JSSC15 [34]
Tech.	**28 nm CMOS**	0.130 μm SiGe	0.130 μm SiGe	0.130 μm SiGe	40 nm CMOS	28 nm CMOS	28 nm CMOS	28 nm CMOS	40 nm CMOS	65 nm CMOS	28 nm CMOS
V_{DD} (V)	**0.9**	1.5	4	4	1.1	1	1.1	2.2	1.8	1	1
f_c (GHz)	**43**	31.5	20–50*	40–65*	27	29	28	28	63	66	53
Gain (dB)	**20.8**	$17^{\sharp}$	23.4*	15.5–23*$^{\sharp}$	22.4	15.7	10	13.6	22.4	30	13
S_{21} $BW_{-3\,dB}$	**65%**	52%	NA	NA	8%$^{\sharp}$	13%	13%$^{\sharp}$	13%$^{\sharp}$	13%	23%	51%
P_{sat} (dBm)	**16.6**	17.1	23.7	23.6	15.1	14	14.8	19.8	16.4	20	13.3
P_{sat} $BW_{-1\,dB}$	**56.6%**	40%*	50%*	47.6%*	NA	10%$^{\sharp}$	NA	NA	12.4%$^{\sharp}$	NA	29.6%
$P_{1\,dB}$ (dBm)	**13.4**	15.5	NA	19.9	13.7	13.2	14	18.6	13.9	16	12
$P_{1\,dB}$ $BW_{-1\,dB}$	**32.3%**	NA	NA	47.6%*$^{\sharp}$	NA	NA	NA	NA	7.1%$^{\sharp}$	NA	29.6%
PAE_{MAX} (%)	**24.2**	22.6	28.5	27.7	33.7	35.5	35.5	43.3	23	22	16
PAE_{MAX} $BW_{-1\,dB}$	**39.5%**	NA	50%*	40%*	NA	NA	NA	NA	12%$^{\sharp}$	NA	35.2%$^{\sharp}$
$PAE_{1\,dB}$ (%)	**12.6**	21.6	NA	15.7	31.1	34.3	35.2	41.4	18.9	9.7$^{\sharp}$	14
$PAE_{1\,dB}$ $BW_{-1\,dB}$	**32.3%**	NA	NA	40%*$^{\sharp}$	NA	NA	NA	NA	7.9%$^{\sharp}$	NA	NA
AM-PM (°)	**0.03/-1^{+}***	10$^{\sharp}$	NA	>2.1$^{\nu}$	NA	6	NA	NA	0.2/0.8^{+}	NA	<2.6
Area (mm^2)	**0.160**	1.735	0.960	0.352	0.225	0.155	0.280	0.280	0.081	0.110	0.056

$^{\sharp}$Grahically etimated. $^{+}$In-band best/worst. *The PA is recunfigured for optimal large signal performance over frequency

$^{\nu}$Differently from all the other works in this table, [14] measures AM-PM as the rms phase error of the constellation of modulated-QPSK signals at the PA output. Further, the non-linearity of the test setup in [14] limits the measurements accuracy and AM-PM is evaluated at ≈3 dB backoff from $P_{1\,dB}$

Table 7.2 Comparison with state-of-the-art silicon-based PAs, modulated signal measurements. ©2018 IEEE. Reprinted, with permission, from [19]

Ref.	**This work** [19]			ISSCC17 [32]			RFIC17 [44]		JSSC17 [14]		ISSCC17 [50]	JSSC16 [16]	TMTT16 [3]	TMTT16 [3]	TMTT16 [20]
V_{DD} (V)	**0.9**			1.5			4		4		1.1	1	1.1	2.2	1.8
Tech.	**28 nm CMOS**			0.13 μm SiGe			0.13 μm SiGe		0.13 μm SiGe		40 nm CMOS	28 nm CMOS	28 nm CMOS	28 nm CMOS	40 nm CMOS
Modulated signal	**64-QAM**			64-QAM			16-QAM		64-QAM		64-QAM OFDM	64-QAM OFDM	64-QAM	64-QAM	64-QAM
$f_{carrier}$ (GHz)	**34**			28*	37*	39*	30*	50*	40*	60*	27	30	28.5	28.5	63
RF BW (GHz)	**0.337**	**0.675**	**1.35**	0.675			1.35		0.675		0.8	0.25	0.08	0.08	0.33
Data rate (Gb/s)	**1.5**	**3**	**6**	3			4		3		NA	1.5	<0.48	<0.48	1.5
EVM^{+} (dB)	< **−25**			−23.4*	−26.7*	−25.1*	−18.2*	−19.2*	−21.7*	−21.8*	−25	−25	−27.4	−27.5	−25
$P_{outEVM (dBm)}$ @	**10.1**	**8.9**	**5.9**	9.2*	9.5*	9.3*	16.4*	16.9*	11.7*	12.8*	6.7	4.2	6.8	11	7
PAE @ EVM **(%)**	**5.8**	**4.4**	**2.3**	8.5*$^{\sharp}$	10*$^{\sharp}$	8.5*$^{\sharp}$	12.4*$^{\sharp}$	12.6*$^{\sharp}$	NA	NA	11	9	16.5	17.3	5$^{\sharp}$
ACPR @ **EVM (dBc)**	**−32.1**	**−30.2**	**−36.9**	−28.4*	−30.3*	−29.8*	NA	NA	NA	NA	−29.4	−26.4	NA	NA	−35

$^{\sharp}$Grahically etimated from CW measurements. $^{+}$Normalized to the reference RMS power (EVM≈-SNR(MER)).
*The PA is recunfigured for optimal large signal performance over frequency

trasmission lines used to realize the desired broadband Doherty performance results in >10x larger silicon area when compared with this design. The state-of-the-art frequency reconfigurable 0.13 μm SiGe PAs reported in [14, 44] achieve outstanding measured continuous wave and modulated signal performance over very wide bandwidth, at the cost of added complexity in the power amplifier design and operation.[2] However, these designs do not meet the EVM requirements, despite the 4 V supply.

Finally, the 0.9 V 28 nm bulk CMOS process without ultra-thick top metal option adopted in this work is more favorable to low-cost, high level of integration and high-volume production.

7.3.5 *Appendix I*

The close form expression of the *kQ* product of the 2-port 4th order filter shown in Fig. 7.20a reported in Eq. 7.6 is derived in this appendix. The *kQ* product of a passive linear reciprocal 2-port network can be expressed as [49]

$$kQ = \frac{|Z_{21}|}{\sqrt{R_{11}R_{22} - R_{12}R_{21}}}, \tag{7.13}$$

where

$$\begin{bmatrix} Z_{11} & Z_{12} \\ Z_{21} & Z_{22} \end{bmatrix} = [R] + j[X] = \begin{bmatrix} R_{11} & R_{12} \\ R_{21} & R_{22} \end{bmatrix} + j \begin{bmatrix} X_{11} & X_{12} \\ X_{21} & X_{22} \end{bmatrix}. \tag{7.14}$$

Tanking advantage of the duality law, we can write

$$kQ = \frac{|Y_{21}|}{\sqrt{G_{11}G_{22} - G_{12}G_{21}}}, \tag{7.15}$$

where by inspection of the circuit in Fig. 7.20a

$$\begin{bmatrix} Y_{11} & Y_{12} \\ Y_{21} & Y_{22} \end{bmatrix} = [G] + j[B] = \begin{bmatrix} \frac{1}{R_O} & 0 \\ 0 & \frac{1}{R_L} \end{bmatrix} + \begin{bmatrix} sC_O + \frac{1}{sL_{O,P}(1-k_O^2)} & \frac{k_O}{s\sqrt{L_{O,P}L_{O,S}}(1-k_O^2)} \\ +\frac{k_O}{s\sqrt{L_{O,P}L_{O,S}}(1-k_O^2)} & sC_{PAD} + \frac{1}{sL_{O,S}(1-k_O^2)} \end{bmatrix}. \tag{7.16}$$

From Eqs. 7.15, 7.16 and 7.5, Eq. 7.6 directly follows.

[2] In these works, novel techniques are introduced to reconfigure the passive networks and/or the bias point of the active stages for optimal large signal performance at each frequency of operation.

7.3.6 *Appendix II*

Referring to Fig. 7.20, this appendix derives the close form expression of the impedance presented at the PA output (i.e. Z_{LA}). The admittance matrix of the two-port network Fig. 7.20a can be derived as

$$Y_{11} = s\frac{C_O}{n} + \frac{1}{s\,n\,L_{O,P}(1-k_O^2)}, \tag{7.17}$$

$$Y_{22} = \frac{1}{R_L} + s\,C_{PAD} + \frac{1}{s\,L_{O,S}(1-k_O^2)}, \tag{7.18}$$

$$Y_{21} = Y_{12} = \frac{k_O}{s\,\sqrt{n\,L_{O,P}\,L_{O,S}}(1-k_O^2)}, \tag{7.19}$$

were in this case R_O has been neglected, since we are interested in Z_{LA} (see Fig. 7.20a). The impedance presented at the PA output can now be expressed as [53]

$$Z_{LA} = \frac{Y_{22}}{Y_{11}Y_{22} - Y_{12}Y_{21}}. \tag{7.20}$$

From Eq.s. 7.7, 7.8, 7.17, 7.18, 7.19, 7.20, and carrying out the algebra Eq. 7.9 follows.

7.4 Conclusion

This chapter discussed the fundamentals of power amplifiers for mm-Wave applications. First, the most significant trade-offs and challenges were discussed in Sect. 7.1. Then, in Sect. 7.2 some intricacies of class-AB operation at mm-Wave frequency have been discussed, several AM-PM distortion sources have been highlighted and most popular state-of-the-art solutions have been revised.

Finally, Sect. 7.3 presented a wideband AM-PM compensated class-AB power amplifier suitable for highly integrated 5G phased arrays. Design techniques to realize broadband impedance transformation, power division/combining and phase distortion linearization were discussed in great detail. Second order effects due to practical layout constrains imposed by deep-scaled technologies have been addressed and simple design solutions have been proposed.

Applied to a 0.9 V 28 nm bulk CMOS power amplifier, the presented design techniques allow a measured P_{sat} =15.1 dBm$\pm$1.6 dB and |AM-PM| $< 1°$ from 29 to 57 GHz, with a peak PAE of 24.2%. When a 64-QAM signal with wide modulation bandwidth is applied, the realized PA enables up to 10.1, 8.9, 5.9 dBm average P_{OUT} while amplifying a 1.5, 3, 6 Gb/s respectively at 34 GHz. The in-band and out-of-band linearity measured in EVM and ACPR is always better than −25 dB and −30 dBc respectively, without any digital pre-distortion.

References

1. S.C. Cripps, Advanced Techniques in RF Power Amplifier Design (Artech House, 2002)
2. S. Cripps, RF Power Amplifiers for Wireless Communications (Artech House, 2006)
3. B. Park, S. Jin, D. Jeong, J. Kim, Y. Cho, K. Moon, B. Kim, Highly linear mm-wave CMOS power amplifier. IEEE Trans. Microwave Theory Tech. **64**(12), 4535–4544 (2016)
4. Y. Zhang, P. Reynaert, A high-efficiency linear power amplifier for 28GHz mobile communications in 40nm CMOS, in *IEEE Radio Frequency Integrated Circuits Symposium (RFIC)* (Honolulu, HI, 2017), pp. 33–36
5. H. Zhang, E. Sanchez-Sinencio, Linearization techniques for CMOS low noise amplifiers: a tutorial. IEEE Trans. Circuits Syst. I: Regul. Pap. **58**(1), 22–36 (2011)
6. M. Shahmohammadi, M. Babaie, R.B. Staszewski, A 1/f noise upconversion reduction technique for voltage-biased RF CMOS oscillators. IEEE J. Solid-State Circuits **51**(11), 2610–2624 (2016)
7. E. McCune, A Technical Foundation for RF CMOS Power Amplifiers: Part 2: Power Amplifier Architectures. IEEE Solid-State Circuits Mag. **7**(4), 75–82 (2015)
8. E. McCune, Fundamentals of switching RF power amplifiers. IEEE Microwave Wirel. Compon. Lett. **25**(12), 838–840 (2015)
9. M. Babaie, R.B. Staszewski, L. Galatro, M. Spirito, A wideband 60GHz class-E/F2 power amplifier in 40nm CMOS, in *2015 IEEE Radio Frequency Integrated Circuits Symposium* (2015), pp. 215–218
10. E. McCune, A technical foundation for RF CMOS power amplifiers: Part 5: making a switch-mode power amplifier. IEEE Solid-State Circuits Mag. **8**(3), 57–62 (2016)
11. S.Y. Mortazavi, K.J. Koh, Integrated inverse class-F silicon power amplifiers for high power efficiency at microwave and mm-wave. IEEE J. Solid-State Circuits **51**(10), 2420–2434 (2016)
12. P. Reynaert, W. Steyaert, M. Vigilante, "RF CMOS." Nanoelectronics: Materials, Devices, Applications, 2 Volumes (2017)
13. D. Zhao, P. Reynaert, A 60-GHz dual-mode class AB power amplifier in 40-nm CMOS. IEEE J. Solid-State Circuits **48**(10), 2323–2337 (2013)
14. C.R. Chappidi, K. Sengupta, Frequency reconfigurable mm-wave power amplifier with active impedance synthesis in an asymmetrical non-isolated combiner: analysis and design. IEEE J. Solid-State Circuits **PP**(99), 1–12 (2017)
15. D. Zhao, P. Reynaert, An E-band power amplifier with broadband parallel-series power combiner in 40-nm CMOS. IEEE Trans. Microwave Theory Tech. **63**(2), 683–690 (2015)
16. S. Shakib, H.C. Park, J. Dunworth, V. Aparin, K. Entesari, A highly efficient and linear power amplifier for 28-GHz 5G phased array radios in 28-nm CMOS. IEEE J. Solid-State Circuits **51**(12), 3020–3036 (2016)
17. Behzad Razavi, *RF Microelectronics*, 2nd edn. (Prentice Hall, New Jersey, 2011)
18. S. Golara, S. Moloudi, A.A. Abidi, Processes of AM-PM distortion in large-signal single-FET amplifiers. IEEE Trans. Circuits Syst. I: Regul. Pap. **64**(2), 245–260 (2017)
19. M. Vigilante, P. Reynaert, A wideband class-AB power amplifier with 29–57-GHz AM-PM compensation in 0.9-V 28-nm bulk CMOS. IEEE J. Solid-State Circuits **53**(5), 1–14 (2018)
20. S. Kulkarni, P. Reynaert, A 60-GHz power amplifier with AM-PM distortion cancellation in 40-nm CMOS. IEEE Trans. Microwave Theory Tech. **64**(7), 2284–2291 (2016)

21. C. Wang, M. Vaidyanathan, L.E. Larson, A capacitance-compensation technique for improved linearity in CMOS class AB power amplifiers. IEEE J. Solid-State Circuits **39**(11), 1927–1937 (2004)
22. W. Ye, K. Ma, K.S. Yeo, 2.5 A 2-to-6GHz class-AB power amplifier with 28.4% PAE in 65nm CMOS supporting 256QAM, in *2015 IEEE International Solid-State Circuits Conference - (ISSCC) Digest of Technical Papers* (San Francisco, CA, 2015), pp. 1–3
23. W.M.C. Sansen, *Analog Design Essentials*, vol. 859 (Springer Science & Business Media, Berlin, 2007)
24. D. Murphy, H. Darabi, A. Abidi, A.A. Hafez, A. Mirzaei, M. Mikhemar, M.C.F. Chang, A blocker-tolerant, noise-cancelling receiver suitable for wideband wireless applications. IEEE J. Solid-State Circuits **47**(12), 2943–2963 (2012)
25. J.A. Jayamon, J.F. Buckwalter, P.M. Asbeck, A PMOS mm-wave power amplifier at 77GHz with 90mW output power and 24% efficiency. IEEE Radio Freq. Integr. Circuits Symp. **2016**, 262–265 (2016)
26. L. Fanori, P. Andreani, A high-swing complementary class-C VCO, in *2013 Proceedings of the ESSCIRC (ESSCIRC)* (2013), pp. 407–410
27. Y. He, L. Li, P. Reynaert, 60 GHz power amplifier with distributed active transformer and local feedback, in *2010 Proceedings of the ESSCIRC (ESSCIRC)* (2010), pp. 4314–317
28. T.H. Lee, *The Design of CMOS Radio-Frequency Integrated Circuits* (Cambridge university press, Cambridge, 2003)
29. D. Kang, B. Park, D. Kim, J. Kim, Y. Cho, B. Kim, Envelope-tracking CMOS power amplifier module for LTE applications. IEEE Trans. Microwave Theory Tech. **61**(10), 3763–3773 (2013)
30. B. Franois, P. Reynaert, Highly linear fully integrated wideband RF PA for LTE-advanced in 180-nm SOI. IEEE Trans. Microwave Theory Tech. **63**(2), 649–658 (2015)
31. S.A.R. Ahmadi-Mehr, M. Tohidian, R.B. Staszewski, Analysis and design of a multi-core oscillator for ultra-low phase noise. IEEE Trans. Circuits Syst. I: Regul. Pap. **63**(4), 529–539 (2016)
32. S. Hu, F. Wang, H. Wang, A 28GHz, 37GHz, 39GHz multiband linear doherty power amplifier for 5G massive MIMO applications, in *2017 IEEE International Solid-State Circuits Conference - (ISSCC) Digest of Technical Papers* (San Francisco, CA, 2017), pp. 1–3
33. M. Vigilante, P. Reynaert, On the design of wideband transformer-based fourth order matching networks for E-band receivers in 28-nm CMOS. IEEE J. Solid-State Circuits **52**(8), 2071–2082 (2017)
34. M. Bassi, J. Zhao, A. Bevilacqua, A. Ghilioni, A. Mazzanti, F. Svelto, A 40–67GHz power amplifier with 13dBm P_{SAT} and 16% PAE in 28nm CMOS LP. IEEE J. Solid-State Circuits **50**(7), 1618–1628 (2015)
35. I. Aoki, S.D. Kee, D.B. Rutledge, A. Hajimiri, Fully integrated CMOS power amplifier design using the distributed active-transformer architecture. IEEE J. Solid-State Circuits **37**(3), 371–383 (2002)
36. P. Haldi, D. Chowdhury, P. Reynaert, G. Liu, A.M. Niknejad, A 5.8GHz 1V linear power amplifier using a novel on-chip transformer power combiner in standard 90nm CMOS. IEEE J. Solid-State Circuits **43**(5), 1054–1063 (2008)
37. E. Kaymaksut, P. Reynaert, Transformer-based uneven doherty power amplifier in 90nm CMOS for WLAN applications. IEEE J. Solid-State Circuits **47**(7), 1659–1671 (2012)
38. J. Zhao, M. Bassi, A. Mazzanti, F. Svelto, A 15 GHz-bandwidth 20dBm PSAT power amplifier with 22% PAE in 65nm CMOS, in *2015 IEEE Custom Integrated Circuits Conference (CICC)* (San Jose, CA, 2015) pp. 1–4
39. H. Wang, C. Sideris, A. Hajimiri, A CMOS broadband power amplifier with a transformer-based high-order output matching network. IEEE J. Solid-State Circuits **45**(12), 2709–2722 (2010)
40. C.R. Chappidi, K. Sengupta, 20.2 A frequency-reconfigurable mm-wave power amplifier with active-impedance synthesis in an asymmetrical non-isolated combiner, in *2016 IEEE International Solid-State Circuits Conference (ISSCC)* (San Francisco, CA, 2016), pp. 344–345

41. E. Kaymaksut, B. Franois, P. Reynaert, Analysis and optimization of transformer-based power combining for back-off efficiency enhancement. IEEE Trans. Circuits Syst. I: Regul. Pap. **60**(4), 825–835 (2013)
42. E. Kaymaksut, P. Reynaert, Dual-mode CMOS doherty LTE power amplifier with symmetric hybrid transformer. IEEE J. Solid-State Circuits **50**(9), 1974–1987 (2015)
43. M. Ozen, K. Andersson, C. Fager, Symmetrical doherty power amplifier with extended efficiency range. IEEE Trans. Microwave Theory Tech. **64**(4), 1273–1284 (2016)
44. C.R. Chappidi, K. Sengupta, Globally optimal matching networks with lossy passivesand efficiency bounds, in *IEEE Radio Frequency Integrated Circuits Symposium (RFIC)* (Honolulu, HI, 2017), pp. 328–331
45. F.-W. Kuo et al., A 12mW all-digital PLL based on class-F DCO for 4G phones in 28nm CMOS, in *2014 Symposium on VLSI Circuits Digest of Technical Papers* (Honolulu, HI, 2014), pp. 1–2
46. C.R. Chappidi, K. Sengupta, Globally optimal matching networks with lossy passives and efficiency bounds. IEEE Trans. Circuits Syst. I: Regul. Pap. **PP**(99), 1–12 (2017)
47. J.R. Long, Monolithic transformers for silicon RF IC design. IEEE J. Solid-State Circuits **35**(9), 1368–1382 (2000)
48. I. Aoki, S.D. Kee, D.B. Rutledge, A. Hajimiri, Distributed active transformer-a new power-combining and impedance-transformation technique. IEEE Trans. Microwave Theory Tech. **50**(1), 316–331 (2002)
49. T. Ohira, The kQ product as viewed by an analog circuit engineer. IEEE Circuits Syst. Mag. **17**(1), 27–32 (2017)
50. S. Shakib, H.C. Park, J. Dunworth, V. Aparin, K. Entesari, A wideband 28GHz power amplifier supporting 8x100MHz carrier aggregation for 5G in 40nm CMOS, in *2017 IEEE International Solid-State Circuits Conference - (ISSCC) Digest of Technical Papers* (San Francisco, CA, 2017), pp. 1–3
51. M. Vigilante, P. Reynaert, A 29-to-57GHz AM-PM compensated class-AB power amplifier for 5G phased arrays in 0.9V 28nm bulk CMOS, in *2017 IEEE Radio Frequency Integrated Circuits Symposium (RFIC)* (Honolulu, HI, 2017), pp. 116–119
52. M. Vigilante, E. McCune, P. Reynaert, To EVM or two EVMs?: an answer to the question. IEEE Solid-State Circuits Mag. **9**(3), 36–39 (2017)
53. D.M. Pozar, *Microwave Engineering* (Wiley, New York, 2009)

Chapter 8
Conclusion

8.1 Summary

A new era is fast approaching. Up to hundred devices and sensors will surround every person, spanning from simple low cost disposable sensors, to smart watches and wearables, from car radar for adaptive cruise control, blind spot detection, etc. to self driving car, not to mention high quality video applications for smartphones, tablet and 360° virtual reality. To enable this revolution $100\times$ higher data rate, $100\times$ higher network efficiency and better than 1ms latency are needed. To ensure low cost and mass production capabilities CMOS technology will play a key role. Therefore, design techniques for broadband and low power building blocks for mm-Wave transceivers integrated in deep-scaled CMOS are attracting an ever increasing attention from industries and research institutes.

This work presents several state-of-the-art mm-Wave building blocks for future 5G TRXs. An introduction on the topic and motivations are given in Chap. 1. The key aspects of mm-Wave active and passive devices implemented in deep-scaled CMOS are given in Chap. 2, while a detailed discussion of on-chip broadband 4th order filters is reported in Chap. 3. The remaining Chapters discuss different building blocks for PLLs, RXs and TXs. First the basics are introduced and the challenges due to the high speed of operation are discussed. Then, the design, layout and measurements of state-of-the-art test chips are discussed. Chapters 4 and 5 are devoted the mm-Wave front-end of a fundamental quadrature PLL. Design techniques for low power, low area, low noise and wide bandwidth of operation are demonstrated. Chapter 6 focuses on low-noise amplifiers and downconverters for E-Band point-to-point communication links. The design techniques discussed in the previous Chapters are leveraged to demonstrate the first broadband highly sensitive receiver that covers the 71–76 GHz and 81–86 GHz frequency bands with wide margin. Such broadband performance is key to account for model inaccuracy and PVT variations, particularly significant at mm-Wave. Power amplifiers are the object of Chap. 7. Several linearization techniques are discussed and a highly linear PA with 65% fractional BW for 5G phased

M. Vigilante and P. Reynaert, *5G and E-Band Communication Circuits in Deep-Scaled CMOS*, Analog Circuits and Signal Processing,
https://doi.org/10.1007/978-3-319-72646-5_8

arrays is demonstrated. All designs were implemented in a 28 nm CMOS technology without RF thick top metal option.

8.2 Major Contributions

The major original contributions of this work are the following.

The impact of the phase noise at the output of a mm-Wave PLL on the system level requirements is discussed in Chap. 1, Sect. 1.3. This effect is particularly challenging to address, and for this reason often neglected in literature. However, it is shown that the PN performance of state-of-the-art integrated PLLs is still severely limiting the link budget and for high order modulation schemes PN is indeed the major bottleneck.

Different definitions of EVM particularly relevant for 5G power amplifiers are discussed in Chap. 1, Sect. 1.3. Since 5G lacks of a defined standard as of this moment of writing, while several 5G power amplifiers are already published, various misleading comparison tables can be found to date in literature. A 3.7 dB difference in the two most used EVM definitions is theoretically demonstrated when a 64-QAM is transmitted, highlighting the importance of this discussion.

Several state-of-the-art 4th order filters proposed for broadband amplifiers are compared in great detail in Chap. 3, Sect. 3.2. It is theoretically demonstrated that transformer-based coupled resonators perform best and are more favorable to on-chip practical implementation.

Second order effects due to physical layout realization in magnetically coupled resonators often neglected in previous literature are deeply discussed in Chap. 3, Sect. 3.3. Simple close form expressions that shed new insights on these pervasive kinds of filters are derived. It is shown that inverting transformers perform best when a large pass-band frequency response is needed, even if they show a theoretically lower self-resonant frequency when compared to non-inverting ones. The effect of unbalanced capacitive terminations on the filter response is shown. Simple close form expressions to achieve frequency equalization are derived. The effect of on-chip parasitic coupling in multistage amplifiers is shown to be critical at mm-Wave, where the limited gain of the transconductors imposes the use of several amplifying stages. Instead of neglecting it, simple techniques to take advantage of it are introduced. Moreover, simple design techniques to achieve impedance scaling, power division and combining are discussed.

An E-Band quadrature voltage-controlled oscillator implemented in 28 nm CMOS is demonstrated in Chap. 4, Sect. 4.3. Two fundamental oscillators are coupled by means of gate-to-drain transformers to realize accurate quadrature phases and switched coupled inductors are added for tuning extension. Closed-form expressions of the oscillation frequency and the tuning extension design parameters are derived. The time-variant nature of the circuit-noise to phase-noise of the presented topology is investigated, resulting in simple guidelines for optimal design. Based on the proposed techniques, the realized prototype is tunable over two bands of almost 5 GHz each separated in frequency, while occupying only 0.031 mm^2. The peak measured phase noise at 10 MHz offset is −117.7 dBc/Hz from a 72.7 GHz carrier and −110 dBc/Hz from a 88.2 GHz carrier and varies less than 3.5 dB within each band.

The design and realization of a wideband tunable divide-by-4 in 28 nm bulk CMOS is presented in Chap. 5, Sect. 5.3. A systematic design methodology to maximize the locking range over power consumption ratio is proposed. The test chip core area is only $25.6 \times 24.8\,\mu m^2$ and measurements repeated over several samples demonstrate an operating frequency range from 25 to 102 GHz with a maximum power consumption of 5.64 mW from a 0.9 V supply. The frequency band from 44.3 to 90 GHz is covered in only three steps with a minimum fractional bandwidth in exceed of 20% and power consumption less than 4.7 mW demonstrating the effectiveness of the proposed design techniques. This is the first time that a single low power divide-by-4 circuit is demonstrated with wide margin over the whole E-Band (60–90 GHz) and beyond.

The design and measurements of a broadband 28 nm bulk CMOS LNA and a sliding-IF receiver tailored for E-Band (i.e. 71–76 GHz and 81–86 GHz) point-to-point communication links are presented in Chap. 6, Sects. 6.4 and 6.5. Leveraging the proposed design methodologies, the E-Band LNA achieves a figure of merit $\approx$10.5 dB better that state-of-the-art designs in the same band and comparable to LNAs at lower frequencies. The RX achieves 30.8 dB conversion gain with <1 dB in-band ripple over a 27.5 GHz $BW_{-3\,\text{dB}}$ while demonstrating a 7.3 dB minimum NF with less than 2 dB variation from 61.4 to 88.9 GHz. The worst cases in-band $ICP_{-1\,\text{dB}}$ and IIP3 are −30.7 dBm and −23.8 dBm respectively from a 0.9 V power supply. This wideband state-of-the-art performance enables robust and low power multi-Gb/s wireless communication over short to medium distance over the complete E-Band with wide margin.

A 29–57 GHz (65% BW) AM-PM compensated class-AB power amplifier tailored for 5G phased arrays is demonstrated in Chap. 7, Sect. 7.3. Designed in 0.9 V 28 nm CMOS without RF thick top metal, the PA achieves a P_{sat}=15.1 dBm ± 1.6 dB and |AM-PM| < 1° from 29 to 57 GHz, with a peak PAE of 24.2%. Techniques are studied to realize the required load impedance and distortion cancellation over the wide band of operation, while allowing 2-way power combining to further increase the delivered P_{OUT}. The very low AM-PM distortion of the realized PA enables up to 10.1, 8.9, 5.9 dBm average P_{OUT} while amplifying a 1.5, 3, 6 Gb/s 64-QAM respectively at 34 GHz with EVM/ACPR better than −25 dB/−30 dBc, without any digital pre-distortion.

8.3 Suggestions for Future Work

Some suggestions for future work both at circuit level and architectural level are summarized in the following.

On oscillator design. In Chap. 4, Sect. 4.1 a general result on phase noise has been introduced. It has been shown that biasing the negative Gm stage in class-C permits to achieve very low power consumption for a given PN [1, 2]. Moreover, it has been shown that due to the limited quality factor of on-chip transmission lines, the tank of a distributed mm-Wave oscillator in a practical implementation also shows only one resonant peak at the fundamental frequency, see Fig. 4.9 [3]. These considerations lead us to expect that a class-C standing wave oscillator and a class-C rotary traveling wave oscillator could be demonstrated. Especially when a deep-scaled technology node is used, where the f_t is remarkably high even at very low biasing currents, as discussed in Chap. 2.

On high speed divider design. Several divider architectures for mm-Wave applications have been already studied in great detail and excellent implementations can be found to date in literature.

On downconverter design. In [4] a 24 GHz sub-harmonic receiver implemented in 65 nm CMOS has been demonstrated. An on-chip double-quadrature oscillator is designed to provide the four required differential phases, as shown in Fig. 8.1. This idea is not novel, however the implementation deserves attention, and the use of passive mixers in particular. Transistors in a passive mixer work as switches, and switch at $f_{LO} = 2f_{RF}/N_{phases}$. If we consider a subharmonic direct conversion receiver for E-Band applications and imagine to find a way to generate $N_{phases} = 8$ differential phases, the PLL could run at $f_{LO} = 2 \cdot 80\,\text{GHz}/8 = 20\,\text{GHz}$ and the MOS switches may really switch when realized in a deep-scaled technology node. Moreover, as discussed in Chaps. 1 and 4 the phase noise at the output of the PLL is a major issue in such systems, and a multiplication factor of 4 would greatly improve the performance of the PLL and in turns of the full system. Not to mention the power consumption that would be saved for the LO distribution network, frequency multipliers and circuits for I/Q generation. However, we still need to find a way to generate the required phases. As some of you may have already understood while reading this part, the rotary traveling wave oscillator discussed in Chap. 4 would beautifully fit this system. If on the one hand this discussion seems promising, still the practical realization of such system at mm-Wave frequencies is not obvious at all. Passive mixers are extremely lossy at mm-Wave, imposing high gain in the RF path

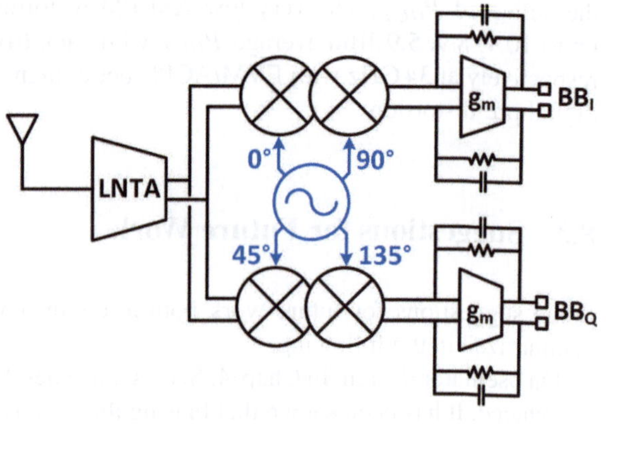

Fig. 8.1 Block diagram of a subharmonic direct-conversion receiver front-end

to keep the noise figure of the full receiver under control. Very high gain in the LNA means very low linearity. Moreover, the phases need to be extracted accurately not to compromise even further the conversion gain of the RX and the I/Q imbalance, posing serious difficulties on the layout.

On power amplifier design. Recently, in [5] a SiGe BiCMOS E-Band power amplifier with a common base stage active device was proposed. The CMOS version would be a common gate stage. Such single transistor amplifier shows higher linearity when compared to the more famous common source stage. This is because it is a non-inverting amplifier, so input and output swing together. However, it also results in lower gain and in this circuit it is not possible to resort to capacitive neutralization. Therefore, the input and output are not well isolated also in differential mode, rising concerns about stability and making the design more involved. Nevertheless, it would be interesting to verify the performance improvement that a G_m-boosted common gate amplifier would have when used as a power amplifier stage. This circuit was successfully adopted at mm-Wave as a first stage of the LNA discussed in Chap. 6, Sect. 6.4, showing promising results. Further, the input impedance of this stage is low, lowering the RC product of the load impedance seen in the inter-stage matching network, improving the linearity of the driver stage under low V_{DD} operation and theoretically improving AM-PM distortion as discussed in Chap. 7, Sect. 7.2. Cascode devices could be added to improve input-output isolation while allowing larger supply voltage, and a complementary N-PMOS realization could further improve linearity and performance under modulated signal [6, 7].

More on power amplifier design. Several works have shown that when a (non isolated) power combiner is driven asymmetrically, it exhibits very interesting properties. One well known example is the Doherty amplifier, where the PAs at the input of the combiner are biased and driven differently allowing excellent improvement in terms of AM-AM distortion and therefore efficiency at power back-off. Some really good references on this topic are [8–14]. However, not much is mentioned about AM-PM distortion (experimental results suggest that is remarkably bad), and in literature a rigorous study that leads to simple design guidelines and intuition on the circuit operation is still missing.

References

1. A. Mazzanti, P. Andreani, Class-C harmonic CMOS VCOs, with a general result on phase noise. IEEE J. Solid-State Circuits **43**(12), 2716–2729 (2008)
2. M. Garampazzi et al., An intuitive analysis of phase noise fundamental limits suitable for benchmarking LC oscillators. IEEE J. Solid-State Circuits **49**(3), 635–645 (2014)
3. A. Moroni, R. Genesi, D. Manstretta, Analysis and design of a 54 GHz distributed hybrid wave oscillator array with quadrature outputs. IEEE J. Solid-State Circuits **49**(5), 1158–1172 (2014)
4. A. Mazzanti, M. Sosio, M. Repossi, F. Svelto, A 24 GHz subharmonic direct conversion receiver in 65 nm CMOS. IEEE Trans. Circuits Syst. I: Regul. Pap. **58**(1), 88–97 (2011)
5. J. Zhao, E. Rahimi, F. Svelto, A. Mazzanti, A SiGe BiCMOS E-band power amplifier with 22% PAE at 18dBm OP_{1dB} and 8.5% at 6dB back-off leveraging current clamping in a common-base

stage, 2017 IEEE International Solid-State Circuits Conference - (ISSCC) Digest of Technical Papers, San Francisco, CA (2017), pp. 1–3
6. I. Fabiano, M. Sosio, A. Liscidini, R. Castello, SAW-less analog front-end receivers for TDD and FDD. IEEE J. Solid-State Circuits **48**(12), 3067–3079 (2013)
7. S. Kulkarni, P. Reynaert, A 60 GHz power amplifier with AM-PM distortion cancellation in 40 nm CMOS. IEEE Trans. Microw. Theory Tech. **64**(7), 2284–2291 (2016)
8. W.H. Doherty, A new high efficiency power amplifier for modulated waves. Proc. Inst. Radio Eng. **24**(9), 1163–1182 (1936)
9. E. Kaymaksut, P. Reynaert, Transformer-based uneven Doherty power amplifier in 90 nm CMOS for WLAN applications. IEEE J. Solid-State Circuits **47**(7), 1659–1671 (2012)
10. E. Kaymaksut, B. Franois, P. Reynaert, Analysis and optimization of transformer-based power combining for back-off efficiency enhancement. IEEE Trans. Circuits Syst. I: Regul. Pap. **60**(4), 825–835 (2013)
11. E. Kaymaksut, P. Reynaert, Dual-mode CMOS Doherty LTE power amplifier with symmetric hybrid transformer. IEEE J. Solid-State Circuits **50**(9), 1974–1987 (2015)
12. M. Zen, K. Andersson, C. Fager, Symmetrical Doherty power amplifier with extended efficiency range. IEEE Trans. Microw. Theory Tech. **64**(4), 1273–1284 (2016)
13. C.R. Chappidi, K. Sengupta, 20.2 A frequency-reconfigurable mm-wave power amplifier with active-impedance synthesis in an asymmetrical non-isolated combiner, IEEE International Solid-State Circuits Conference (ISSCC). San Francisco, CA (2016), pp. 344–345
14. S. Hu, F. Wang, H. Wang, A 28 GHz, 37 GHz, 39 GHz Multiband linear Doherty power amplifier for 5G massive MIMO applications, IEEE International Solid-State Circuits Conference - (ISSCC) Digest of Technical Papers. San Francisco, CA (2017), pp. 1–3

Index

M. Vigilante and P. Reynaert, *5G and E-Band Communication Circuits in Deep-Scaled CMOS*, Analog Circuits and Signal Processing, https://doi.org/10.1007/978-3-319-72646-5

P

Q

R

S

T

U

V

MIX
Papier aus verantwortungsvollen Quellen
Paper from responsible sources
FSC® C105338
FSC
www.fsc.org

If you have any concerns about our products,
you can contact us on
ProductSafety@springernature.com

In case Publisher is established outside the EU,
the EU authorized representative is:
Springer Nature Customer Service Center GmbH
Europaplatz 3, 69115 Heidelberg, Germany

Printed by Libri Plureos GmbH
in Hamburg, Germany